T5-CVD-949

FLOWMETERS

A Comprehensive Survey and Guide To Selection

FLOWMETERS

A Comprehensive Survey and Guide To Selection

Furio Cascetta
Paolo Vigo

FLOWMETERS A Comprehensive Survey and Guide to Selection

Printed in the United States of America

INSTRUMENT SOCIETY OF AMERICA
67 Alexander Drive
P. O. Box 12277
Research Triangle Park
North Carolina 27709
Phone: (919) 549-8411
Telex: 802-540

Library of Congress Cataloging-in-Publication Data

Cascetta, Furio.
Flowmeters : a comprehensive survey and guide to selection.

1. Flow meters-Handbooks, manuals, etc.
I. Vigo, Paolo. II. Title.
TJ935.C39 1988 681′.2 88-8871
ISBN: 1-55617-099-8

Contents

Preface

Instrument specialists must frequently cope with the task of selecting the right flowmeter. The choice is not easy because of: (1) the continuously increasing number of meters available on the market, (2) the various flowmetering installation requirements, and (3) the lack of comparative information on the characteristics of the different flowmeters.

This is a comprehensive survey of industrial flowmeters, based on actual availability in the marketplace. A flowmeter selection chart, based on operating principles, advantages, and limitations and intended as a guide for instrument specialists, is also included.

Part 1
Introduction

Over the last twenty years flow measurement technologies have become applicable in many more fields than ever before. This is due both to the ever growing demand for measurement and control of fluid flow rate in process plants (related to avoiding waste) and to the wide range of applications of flowmeters in non-industrial fields. Some examples of such applications are the supply and sale of natural gases and/or hot water heaters in the home and fluid measurement and control in the field of biomedicine.

For several reasons, it is not feasible to consider one universal flowmeter [1-3]. Many types of fluids, a great variety of plant and fluid-dynamic operating conditions, and various metrological applications characterize each individual flowmeter installation [4,5]. High reliability is required of a flowmeter installed in a nuclear power station for metering hard fluids [6], while an even greater accuracy of performance is required of a flowmeter used in the sale of liquid and gaseous fuels where a careful control of consumption and an adequate guarantee of parties are imperative [7]. Therefore, a wide range of types of flowmeters must be available.

About 50 basic types of flowmeters can be classified, according to their theory of operation, into 9 groups as follows: differential pressure, variable area, positive displacement, inferential, fluid-dynamic, tracer, electromagnetic, ultrasonic, and mass flowmeters [8-11]. Because of the various models and the various measurement principles used (sandwich, insertion, clamp-on, bypass) several hundred models are available. Each model, in turn, is manufactured in sizes suitable for the various pipe diameters and operating flow rates. Over the last 10 years the market has been spreading out because of the demand for more automatic and remote operations and the rapid progress in electronic technology. Furthermore, such technology has fostered the realization of many "new" methods, previously known only from a theoretical point of view, thus aiding in improving the performances of traditional flowmeters and at the same time overcoming their deficiencies and widening their fields of application [12-14]. Therefore, a downright challenge between the "old" and the "new" methods of measurement is being witnessed. The "old" ones such as differential pressure (except for target meters), variable area, inferential, and positive displacement meters are taking up a new life, mostly from the improved performances of their secondary elements, and show long-standing reliability of operation and a high level of standardization [15]. On the other hand, the "new" methods such as tracer, electromagnetic, ultrasonic, fluid-dynamic, target, and mass meters present the great advantage either of being non-intrusive or of having a wider flow range and greater versatility in measuring some specific fluids [16]. On the whole, the "new" methods show higher performance than the "old" ones do.

The most interesting of the emergent methods are mass flowmeters, especially the Coriolis acceleration types for liquids and the thermal ones for gases. Such flowmeters present high accuracy in measuring mass flow rate, which is essential for assessing mass and energy balances and for all fuel sale activities. Therefore, mass flowmeters are now consuming quite a large share of the market pie to the detriment of volumetric flowmeters, which still measure mass flow indirectly (compensation techniques, flow computers, etc.) [17].

The following considerations drawn from the technical literature can help in understanding the main reasons for the present developing trend of the flowmeter market.

(1) Since the early 1960s the market availability of each individual group of flowmeters has almost doubled. Furthermore, within each group, several different designs have been realized [18]. Consider, for instance, (a) the appearance on the market of target and Pitot averaging meters for the differential pressure group; (b) vortex and fluidic oscillator meters (Coanda effect) for the fluid-dynamic group; and (c) the doubled number of measurement methods for the mass flowmeter group. This doubling is essentially due to the realization of the Coriolis acceleration and thermal meters, the number of designs of which, in their turn, have more than tripled over the last five years [19].

(2) The technical literature on flowmeter choice points out that the mid 1970s and early 1980s witnessed the

substitution of the ultrasonic for the variable area in the six most widely used groups known as "the Big Six" [3] or "the six favorites" [20]. This is probably due to the frequent difficulties of transduction and inconstant accuracy of measurement in the flow range on the part of variable area meters, as well as to the non-intrusive principle of operation of ultrasonic flowmeters. It is probable that in time the mass flowmeter group will be ranked seventh in the popularity poll.

(3) The market share of differential pressure meters (target and averaging Pitot meters excepted) has dropped from 80% in the early 1970s [21] to the mid '80s estimate of just a little short of 40% [12].

(4) Market surveys on the US flowmeter scene, carried out in the early '80s by the Venture Development Corporation [22] (Figure 1) and *Chemical Engineering* predicted: (a) defense of share of the market pie on the part of differential pressure devices (target meters excepted); and (b) probable doubling of vortex-shedding sales by 1988, of ultrasonic meter sales by 1988, and of mass flowmeters by 1990. A similar study by Frost & Sullivan, intended for the European market, confirms such trends even though stating a slower pace. Such estimates are proving correct, especially as far as vortex shedding and mass flowmeters are concerned.

The great variety of products available in an ever-changing market as well as the nonfeasibility of a universal flowmeter (or, rather, of one flowmeter that flowmeter [2,14,23]. In fact, the specialist must take into account the great variety of practical measurement situations peculiar to the fluid flow in closed conduit systems, while considering the related total costs with utmost care. The right flowmeter is surely the one that does its job satisfactorily at the lowest possible cost [24,25]. A thorough knowledge of flowmeter state of the art and their operating principles and working behavior is obviously of crucial importance for making a successful choice. Unfortunately, all too often such data are not clearly assessed and are often only inferred from the technical literature. In order to help the specialist to make the right choice, an up-to-date survey of flowmetering methods in present use in closed conduit systems, as well as their main metrological characteristics and operating performances, is included herein. Once the measurement purposes have been clearly defined, the instrumentation specialist can draw a list of "feasible" flowmeters in relation to the properties of the fluid. From this list can be selected another list of "probable" flowmeters, chosen on the basis of the interaction between the meter itself and its installation (see plant and fluid-dynamic conditions) and the optimization of the ratio of total cost to performance, which is the very heart of industrial ethics [26,27].

Finally, the specialist must compare his choice with other users' experiences and vendors' proved statements, which, in similar application, are often the best credentials for a long-term evaluation of flowmeter behavior [5].

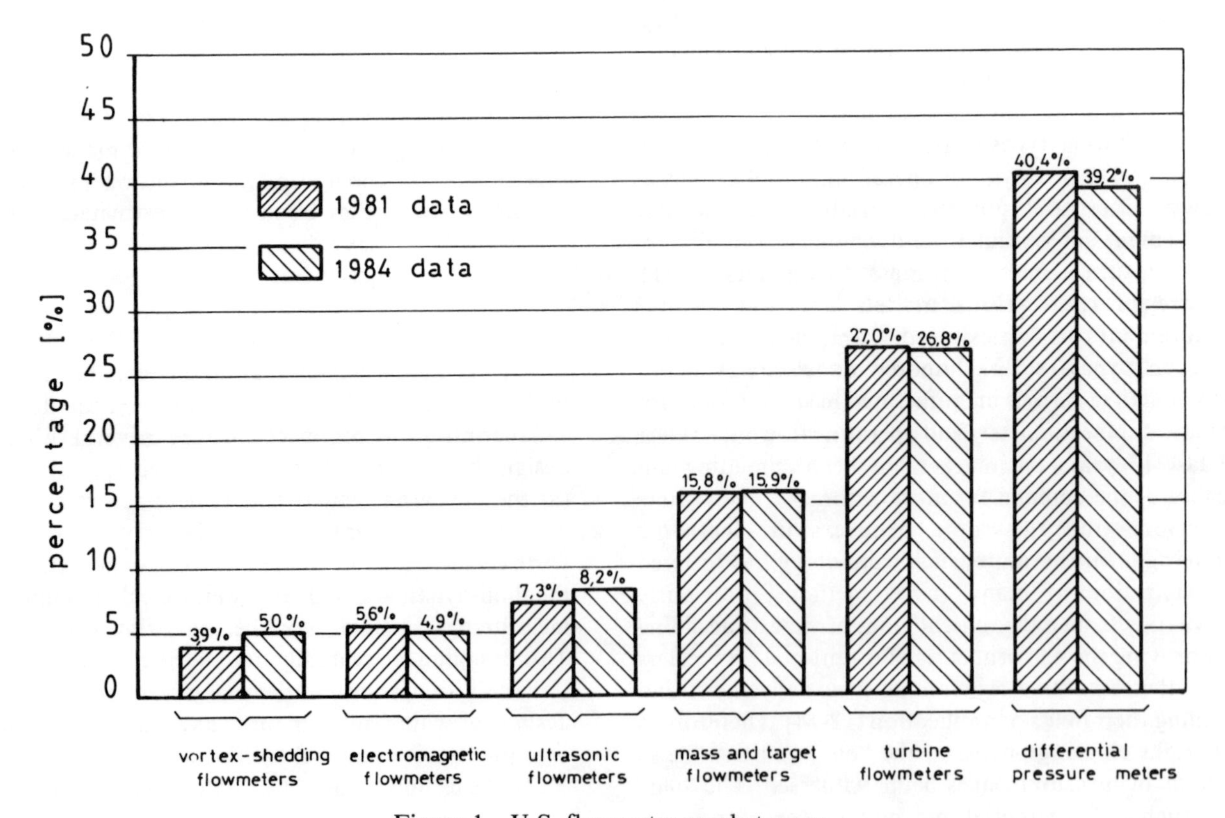

Figure 1—U.S. flowmeter market scene

Part 2
Theory and Operating Principles

This section briefly surveys industrial flowmeters along with their operating principles and theoretical derivations. Data concerning attainable accuracy, flow range, working conditions, advantages, and disadvantages are stated for each meter.

Fluid meters or primary flow elements are divided into 9 groups, as shown in Table 1. The classification is essentially the one adopted in the majority of publications on fluid meters, but for each group only the flowmeters that are widely used and have proved successful in process plants are sketched and described.

The first five groups shown in Table 1 (differential pressure to fluid-dynamic flowmeters plus angular momentum meters from the 9th group) can be, in turn, classified under a wider category known as "energy extractive" [11]. This term implies that all meters in this category exploit the energy of the fluid flow. Furthermore, energy extractive meters, being intrusive, usually involve relevant pressure losses.

The remaining four groups in Table 1 (tracer to mass flowmeters) can be defined as "energy additive" [11] meters, because they introduce to the fluid flow external additional energy in various forms (i.e. electromagnetic, mechanical, or acoustic energy) to operate.

It is worth pointing out that some authors [3, 10] split the 9th group, mass flowmeters, into two sub-classes: "mass flowmeters" and "true mass flowmeters". The term "true mass flowmeters" is used to describe a meter that indicates the mass flow rate of a fluid without the operator being aware of the physical properties of the fluid concerned. The term "mass flowmeter" is sometimes applied to thermal (calorimetric) meters or to other meters that indicate the mass flow rate of a fluid (usually a gas) provided that the nature of the fluid is constant and that its properties are known. Therefore, such meters are not "true" mass flowmeters.

Differential Pressure Flowmeters

Differential pressure (Δp) meters (or head meters) are probably the most widely used for industrial flow measurements. They are available in a great variety of types: orifice, Venturi tube, Dall tube, nozzle, elbow, Pitot tube, averaging Pitot tube, and target.

The orifice (in six types: square edge or concentric, quadrant, conical entrance, eccentric, segmental, integral), the Venturi tube, the Dall tube, and the nozzle are based on the same basic principle—Bernoulli's theorem. The theorem states that fluid to be measured is accelerated when passing through a constriction, thus temporarily increasing the kinetic energy at the expense of the pressure energy. The pressure difference Δp between the pressure taps, which are appropriately located upstream at the full pipe section and downstream of the flow restriction element (primary element) in the proximity of the "vena contracta" section, varies according to the square of volumetric flow rate Q_V:

$$Q_V = CA \sqrt{\frac{\Delta p}{\rho}} \tag{1}$$

where C is the discharge coefficient, ρ is the fluid density, and A is the restricted section area. The relationship in Equation (1) is only apparently simple; behind the practical determination of the discharge coefficient C hides the non-coincidence between the theoretical model and the actual flow motion. That is due mainly to the application of macroscopic balances, based on a one-dimensional hypothesis, to a flow pattern that is strongly characterized by areas of high turbulence and vorticity as well as to the complex connection between the theoretical and the real positions of pressure taps in each device [28-30].

The presence of a flow restriction element in head-type flowmeters causes pressure losses. The less the primary element is devoted to guiding flow (particularly downstream of the contraction where pressure recovery occurs), the higher the pressure losses produced.

Another operating limitation of head-type flowmeters results from the square root relationship of Equation (1), which restricts their rangeability typically to 4:1.

The advantages of these flowmeters lie in their international standardization [31] (even if not coincident among each different standard organization: AGA, ANSI, API, etc.), their simplicity of use, their low cost (particularly for orifice meters), the absence of moving parts, and their ability to operate in pure fluids for a certain time without requiring frequent maintenance or recalibration.

FLOWMETER GROUPS	PRESSURE		
DENOMINATION	O	V	N
SKETCH			
MEASURED QUANTITY	$\Delta p = p_I - p_{II}$	$\Delta p = p_I - p_{II}$	$\Delta p = p_I - p_{II}$
FLUID SERVICE	L, G, V	L, G, V	L, G, V
MAX. TEMPERATURE [°C]	500	500	500
MAX. PRESSURE [MPa]	60	60	60
MIN. REYNOLDS-NUMBER	$5 \cdot 10^3$ (1)	$2 \cdot 10^5$	10^4
MAX. KINEMATIC VISCOSITY [m^2/s]	10^{-2}	10^{-2}	10^{-2}
ACCURACY [%]	± 2 F.S. (2)	± 2 F.S.	± 2 F.S.
REPEATABILITY [%]	±0.25	±0.25	±0.25
SCALE	Sq	Sq	Sq
RANGEABILITY	4/1	4/1	4/1
PRESSURE LOSSES	H	L	M (H)
EASE OF INSTALLATION	E	FE	FE
UPSTREAM STRAIGHT PIPE [L/D]	10 ÷ 80	5 ÷ 30	10 ÷ 80
PURCHASE COST / INSTALLATION COST	L / M	H / M	H / M
OPERATING COST / MAINTENANCE COST	H / L	L / M	M / M
TOTAL COSTS/PERFORMANCE	M	M	H
MANUFACTURERS (*)	m	m	m

DIFFERENTIAL	E	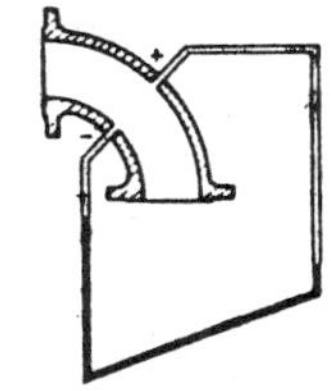	$\Delta p = p_I - p_{II}$	L,G,V	500	60	$5 \cdot 10^4$	-	(3) ±4	±0.50	Sq	4/1	L	E	-	L / L	L / L	M	(8) vfw
	P	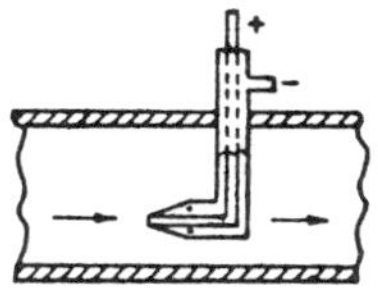	$\Delta p = p_t - p_s$	L,G,V	500	20	10^4	10^{-2}	(4) ±2 F.S.	±0.50	Sq	4/1	L	VE	50	L / L	L / L	M	m
	AP	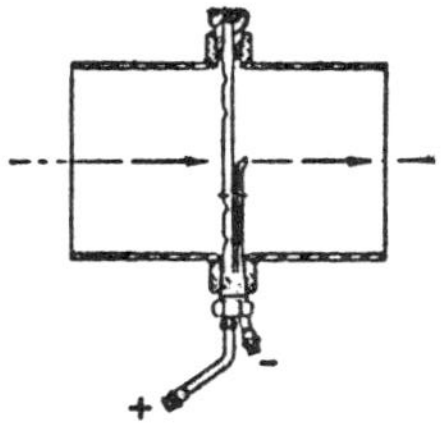	$\Delta p = p_t - p_s$	L,G,V	500	20	10^3	10^{-2}	±1.5 F.S.	±0.50	Sq	4/1	L	VE	10	L / L	L / L	L	fw
	TR	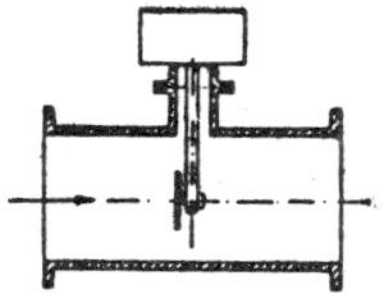	$F_i = c_d \varrho \frac{v^2}{2} A$ $p_t - p_s = \frac{F_i}{c_d A}$	L,G (V)	300	60	$5 \cdot 10^2$	10^{-3}	(5) ±3 F.S. (6) ±1 F.S.	±0.25	Sq	4/1 ÷ 10/1	H	E	10 ÷ 80	L / H	L / H	M	fw
	L		$\Delta p = p_I - p_{II}$	G					(7)		Li	10 ÷ 1		NE					(8) vfw

(*) based on data in "ISA Directory of Instrumentation" 1985-86

TABLE 1

FLOWMETER GROUPS	VARIABLE AREA		DISPLACEMENT
DENOMINATION	R	CP	RV
SKETCH			
MEASURED QUANTITY	z	z	n
FLUID SERVICE	L,G,V	L	L,G
MAX. TEMPERATURE [°C]	(9) 200 / (10) 400	400	(12) 250 / (13) 50
MAX. PRESSURE [MPa]	(9) 4 / (10) 40	60	20
MIN. REYNOLDS-NUMBER	10^4	-	-
MAX. KINEMATIC VISCOSITY [m^2/s]	$5 \cdot 10^{-4}$	10^{-2}	$3 \cdot 10^{-1}$
ACCURACY [%]	(11) ±2 F.S.	±2 F.S.	±0.5
REPEATABILITY [%]	±0.50	±0.50	±0.05
SCALE	Li	Li	Li
RANGEABILITY	10/1	10/1	(12) 10/1 / (13) 20/1
PRESSURE LOSSES	L	H	M
EASE OF INSTALLATION	E	FE	FE
UPSTREAM STRAIGHT PIPE [L/D]	-	-	-
PURCHASE COST / INSTALLATION COST	L / L	M / L	H / L
OPERATING COST / MAINTENANCE COST	L / L	M / L	M / H
TOTAL COSTS/PERFORMANCE	L	M	L
MANUFACTURERS (*)	m	fw	m

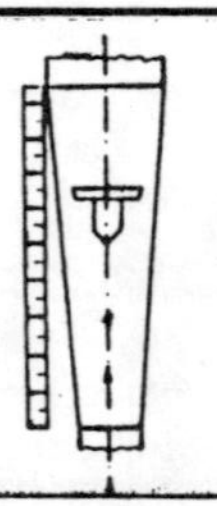

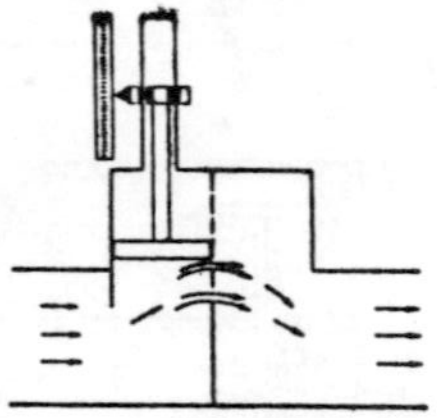

POSITIVE	ND	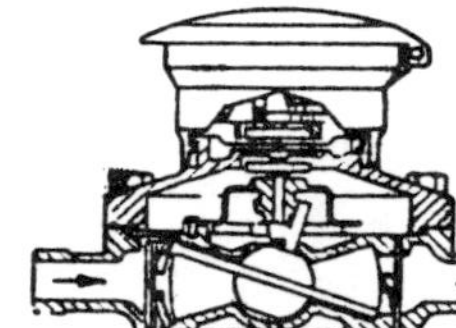	n	L,G	(12) 250 / (13) 50	20	–	$3 \cdot 10^{-1}$	±0.5	±0.05	Li	(12) 10/1 / (13) 20/1	M	FE	–	L / H	H / M	L	m
INFERENTIAL	TU	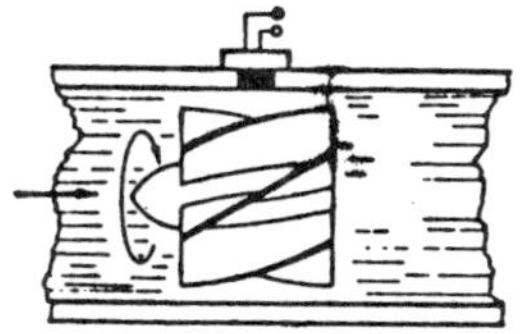	n	L,G	150	40	10^{4}	$3 \cdot 10^{-4}$	±0.5	±0.05	Li	(14) 10/1 ÷ 100/1	L	E	10	M / H	M / L	L	m
FLUID - DYNAMIC	VXS	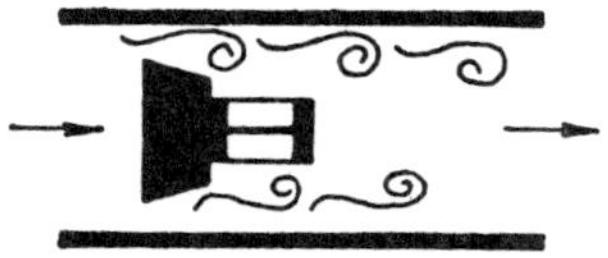	f	L,G,V	400	10	10^{4}	10^{-4}	(12) ±1 / (13) ±2	±0.15	Li	(12) 20/1 / (13) 100/1	L	E	20	L / M	L / L	L	m
	VXP	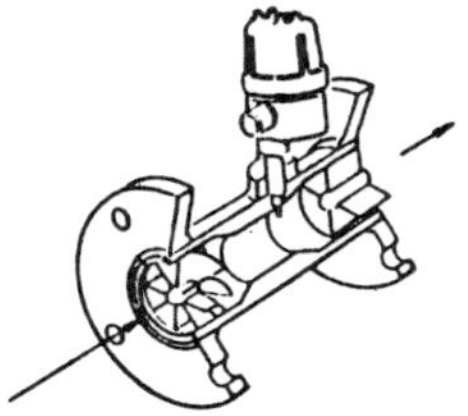	f	G,(L)	200	40	10^{5}	-	±1	±0.25	Li	100/1	H	E	20	L / H	L / H	M	fw
	FO	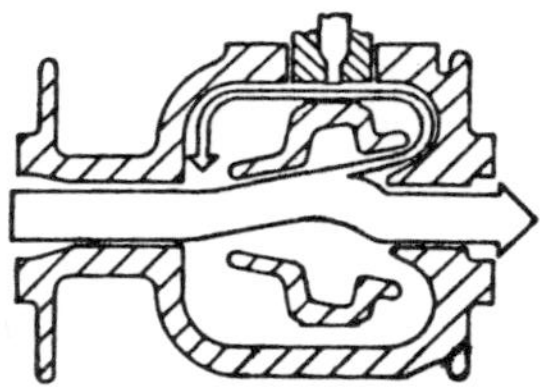	f	L	200	10	$3 \cdot 10^{3}$	10^{-4}	±2 F.S.	±0.20	Li	20/1	M	FE	20 ÷ 50	L / M	M / M	M	vfw

TABLE 1 (Continued)

FLOWMETER GROUPS	DENOMINATION	SKETCH	MEASURED QUANTITY	FLUID SERVICE	MAX. TEMPERATURE [°C]	MAX. PRESSURE [MPa]	MIN. REYNOLDS-NUMBER	MAX. KINEMATIC VISCOSITY [m^2/s]	ACCURACY [%]	REPEATABILITY [%]	SCALE	RANGEABILITY	PRESSURE LOSSES	EASE OF INSTALLATION	UPSTREAM STRAIGHT PIPE [L/D]	PURCHASE COST / INSTALLATION COST	OPERATING COST / MAINTENANCE COST	TOTAL COSTS/PERFORMANCE	MANUFACTURERS (*)
TRACER	NMR	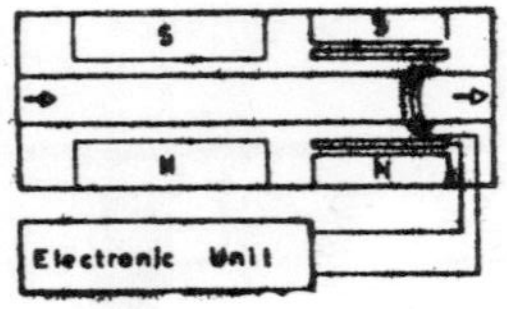	f	(15) L,G,V	-	-	(16) $6 \cdot 10^3$	-	±1	±0.5	Li	10/1	-	NE	-	H / H	H / H	H	vfw
MAGNETIC	EM	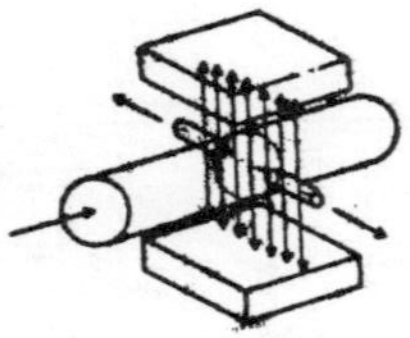	e (mV)	(17) L	200	20	-	-	±0.5	±0.25	Li	10/1	-	E	3 ÷ 5	H / L	L / L	L	m
SONIC	TOF	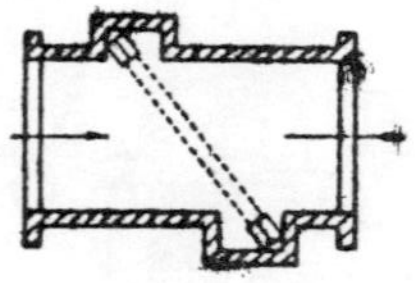	Δf	L,G	250	20	10^4	-	±1	±0.25	Li	20/1	-	E	20	H / L	L / L	M	m

ULTRA	DF	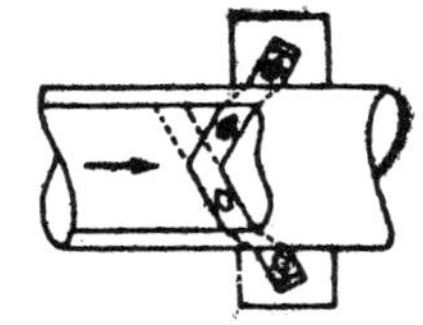	Δf	L,G	250	20	$4\cdot10^{3}$	-	± 2 F.S.	±0.25	Li	20/1	-	E	20	L/H	L/L	M	m
MASS — NOT TRUE	TH	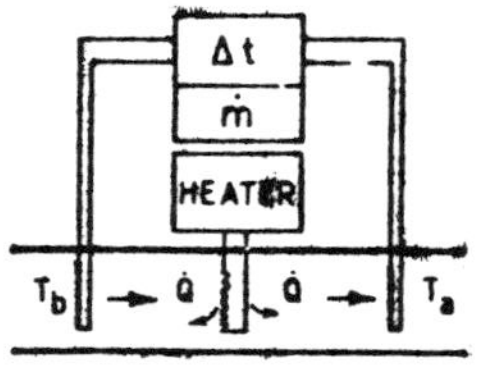	ΔT = $T_{up} - T_{down}$	(18) G,(L)	(18) 50	(18) 1	(20)	-	1÷2 F.S.	±0.25	Log	(18) 30/1	(18) L	(25) FE	(18)	L (22)/H	M/L	M	(23) fw
				(19) G	(19) 200	(19) 10	(21) 10	-				(19) 10/1	(19) L		(19) 10				
MASS — TRUE	WB	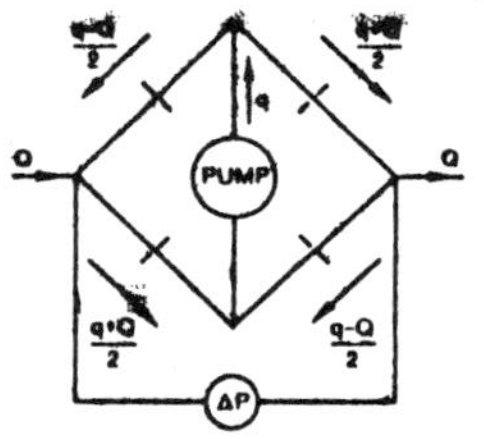	Δp	L	50	1	10^{4}	10^{-2}	±0.5	±0.25	Li	100/1	(24) H	NE	-	M/H	M/H	M	vfw
MASS — TRUE	AM	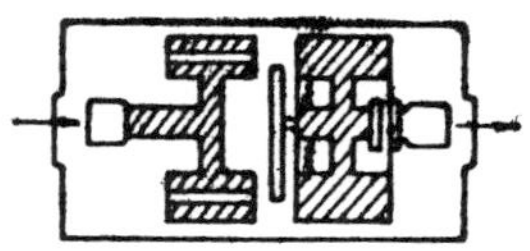	Mt	L,G	400	60	10^{4}	10^{-4}	±0.5	±0.25	Li	10/1	H	FE	10	M/H	M/H	M	vfw
MASS — TRUE	CA	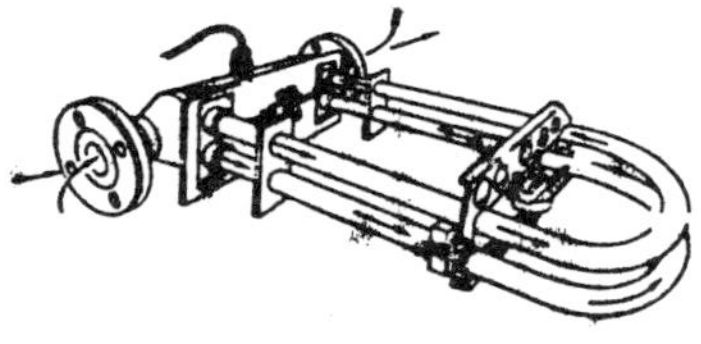	Δt	L,(G)	250	10	-	-	±0.15 ÷ ±0.40	±0.04	Li	20/1 ÷ 100/1	H	E	-	L/H	L/H	L	(23) fw

TABLE 1 (Continued)

ABBREVIATIONS in Table 1

Flowmeter types

O	= orifice	e	= voltage
V	= Venturi tube	f	= frequency
N	= nozzle	fw	= few
E	= elbow meter	m	= many
P	= Pitot tube	n	= rotor velocity
MP	= multiple Pitot tube	p	= pressure
TR	= target meter	t	= time
L	= laminar meter	vfw	= very few
R	= rotameter	z	= height
CP	= cylinder-piston meter	AF	= actual flow rate
RV	= rotary vane meter	E	= easy
ND	= nutating disk meter	FE	= fairly easy
TU	= turbine meter	FS	= full scale
VXS	= vortex-shedding meter	G	= gas
VXP	= vortex precession meter	H	= high
		L	= liquid
FO	= fluidic oscillator meter	Li	= linear
EM	= electromagnetic meter	Log	= logarithmic
NMR	= nuclear magnetic resonance meter	Lw	= low
		M	= medium
TOF	= time of flight meter	Mt	= torque
DF	= Doppler flowmeter	NE	= not easy
TH	= thermal meter	Sq	= square root
WB	= Wheatstone bridge meter	T	= temperature
		V	= vapor
AM	= angular momentum meter	VE	= very easy
CA	= Coriolis acceleration flowmeter		

NOTES in Table 1

(1) For conical entrance orifice $Re < 5 \cdot 10^3$

(2) When properly installed ± 0, 75% of rate

(3) For calibrated meters the accuracy should be ±2% FS

(4) If carefully mounted (ISO 3966, 7145)

(5) In laminar regime

(6) In turbulent regime

(7) Very high, if all the hypothesis of Poiseuille-Hagen solution are respected

(8) User often coincides with manufacturer

(9) Glass tapered tube

(10) Metal tapered tube

(11) After calibration accuracy reaches ± 0, 5% FS

(12) Liquid service

(13) Gas service

(14) Depending upon fluid viscosity

(15) Fluids containing hydrogen or fluorine compounds with magnetic moment

(16) Only for laminar or fully turbulent regimes

(17) Electrical conductivity greater than 5 S/cm

(18) In meters with capillary tube meters

(19) Insertion-type flowmeters

(20) The flow regime must be laminar both in main and in capillary tube

(21) From ISO 7145

(22) High installation costs instead for capillary tube meters

(23) But recently in rapid expansion

(24) Pressure losses are usually considered low, because the energy consumption due to the pump is not taken into account

(25) Very easy for insertion models

For correct installation, all head-type flowmeters require the presence of straight pipes upstream of the primary element and, therefore, a fully developed, non-swirling, symmetrical turbulent velocity profile upstream of the restriction. The installation requirements, the need to reduce permanent pressure losses when using head meters, and the specific requirements for metering particular industrial fluids can induce the specialist to use differential pressure primary devices other than orifice, nozzle, and Venturi tubes. Such alternative devices (elbow meters, target meters, Pitot and averaging Pitot tubes) are known as "nonconventional" head-class primary elements.

The elbow flowmeter is also called the centrifugal meter because of the centrifugal force operating in it. In fact, as a fluid passes through a pipe elbow, the pressure at the outside radius of the elbow itself increases because of the effect of the centrifugal force. If pressure taps are located at the outside and inside of the elbow at 45°, a measurement can be made by metering the pressure difference between the taps. Also in this case the volumetric flow rate is expressed by a square-root relationship, typical of differential pressure devices. Centrifugal meters are characterized by low pressure losses and by low installation costs since they exploit part of the plant piping system already in existence [32]. The major disadvantages of elbow meters are the very low differential pressure created, especially for gas flows, and their low accuracy (±4%) for uncalibrated meters. For flow-calibrated elbow meters, accuracy should be comparable to that available with conventional head-class flowmeters (such as orifices, nozzles, and Venturi tubes).

Pitot tubes and averaging Pitot tubes are also head-class flowmeters. These devices consist of appropriately arranged pressure taps in a cross-flow probe. The taps are located strategically so as to sense the total or impact pressure, P_t, and the static pressure, P_s, respectively. The pressure difference, as shown by Bernoulli's equation for an incompressible fluid motion (i.e., Mach number $M < 0.3$), is proportional to the square root of the local flow velocity:

$$p_t - p_s = \frac{1}{2}\, \delta\, V^2 \rightarrow V = \sqrt{\frac{2(p_t - p_s)}{\rho}} \qquad (2)$$

The average fluid velocity can be inferred from the local velocity and allows the assessment of the volumetric flow rate Q_v. In Pitot tubes the average velocity can be inferred from the velocity profile, determined in turn after a series of local velocity measurements [33], or, more easily but less accurately, by positioning the probe at a particular point ("at a distance of 0.242 R, R being the radius of the conduit, from the wall of the conduit", ISO 7145-1982) where local velocity and average velocity are essentially the same. The Pitot tube is also called "insertion" meter. It is an inexpensive method for measurement of clean liquids, gases, or vapors in large-size pipes [34].

In averaging Pitot tubes the measure of average velocity is obtained by positioning on the Pitot probe, which is inserted across the diameter of the conduit, four or more total pressure taps at mathematically defined positions (Tchebicheff method). Such taps average the total pressures for axisymmetric velocity profiles; hence, its name. It should be noted that in Annubar® meters, unlike in all other types of Pitot tubes, the static pressure holes are located behind the averaging tube so as to sense the negative pressure of its wake. This position of the static pressure holes is intended to significantly amplify the pressure difference to be sensed [35, 36],

All Pitot tubes are convenient for large pipe sizes because of their low cost and their thin shape, which allows easy and rapid mounting, especially in pre-existing plants. Therefore, these devices provide practical and relatively inexpensive measurement with low Δp. The measuring fluid must be a clean liquid, gas, or vapor. The peculiar disadvantages of these meters are related to their being prone to clogging in dirty fluids and to the uncertainties due to non-axisymmetric flow patterns and changes in density. As for all the meters of the differential pressure group, the rangeability in Pitot tubes is limited to 3:1, but it can come to 5:1 for the averaging types.

The latest interesting development in differential pressure devices is the target meter. It can be thought of as an orifice plate turned inside out [3]. In fact, the primary element is a solid circular disk (target) mounted perpendicularly to the flow and suspended at the level of the pipe axis by an electrically controlled force-balance system that measures the impact force F_i on the disk:

$$F_i = c_d\, \rho\, \frac{V^2}{2}\, A \qquad (3)$$

where c_d is a drag coefficient (determined by laboratory tests), ρ is the fluid density, V is the average fluid velocity, and A is the area of the disk. According to Bernoulli's theorem, this force F_i is proportional to the difference between the total pressure and the static pressure, so that the volumetric flow rate, Q_v, can be obtained once again by a square root relationship. Target flowmeters are particularly well-suited for relatively low Reynolds number and dirty flows where the classical head-type flowmeters cannot be used, but they are also used for clean fluids and natural gas. After proper calibration and installation, target meters can provide good accuracy, but their best features are their great rangeability and rapid response in transient testing [37, 38].

As has been said, all the flowmeters previously mentioned operate especially well in turbulent flow regimes, because in this regime the one-dimensional hypothesis is respected. Also included in the differential pressure group is the laminar flowmeter, based as it is on the exact Hagen-Poiseuille solution of the Navier Stokes equations, i.e., the isothermal and fully developed steady flow of a viscous incompressible fluid in a straight circular pipe (Reynolds number < 2000). In order to obtain the measure of the flow rate, the stream is split and driven into a number of parallel capillary tubes where the Poiseuille law is applicable. Therefore, the relationship between pressure difference and volumetric flow rate is as follows:

$$Q_v = \frac{\pi D^4}{128\mu L_i} \Delta p \qquad (4)$$

where L_i is the length of each capillary tube, D is their diameter, and μ is the fluid viscosity. It is worth pointing out that the primary laminar flow element exists in two forms: as capillary tube bundles or a porous plug. Capillary tubes require a set ratio between length and bore greater than 150 so as to minimize inlet and outlet pressure losses [39]. Such meters are rather bulky, expensive, and require careful calibration, and viscosity has a profound effect on the measurement. Furthermore, even though they are suitable for use with gases, they require a high degree of filtration in order not to clog laminar passages. They possess a wide rangeability (10:1) and very high accuracy since, as previously stated, the Poiseuille flow is an exact solution.

A short list of the advantages and limitations of differential pressure flowmeters is given in Table 2.

Table 2

Differential Pressure Flowmeters

Advantages	Limitations
1. Low cost	1. Square root head/flow relationship
2. Easily installed and/or replaced	2. High permanent pressure losses
3. No moving parts	3. Low accuracy
4. Suitable for most gases and liquids	4. Flow range limited to 4:1
5. Available in a wide range of sizes and models	5. Accuracy affected by wear and/or damage of the flow primary element with corrosive fluids

Variable Area Flowmeters

Variable area flowmeters infer the volumetric flow rate, $Q_v = VA$, varying the area A of the flow restriction according to the flow rate, and keeping constant the average fluid velocity V. In other words, while in head-class flowmeters the area A of the restriction is constant, and the average fluid velocity (i.e., the square root of the differential pressure Δp) changes according to the flow rate; in variable area meters the velocity (i.e., the Δp) remains constant, and the area of the restriction varies.

A number of quite dissimilar devices can be classified as "variable area". They all are characterized by an orifice formed by two members, of which one (a tube) is fixed and the other one (a float) is movable. The float and the tube are so arranged that the annular passage between the maximum circumference of the float and the wall of the tapered tube widens as the flow rate increases. The best known type of this group is probably the rotameter, in which the metering fluid flows through a vertical tapered glass, metal, or transparent plastic tube from bottom to top. The tapered tube contains a float (either spherical or cone-shaped), which is drawn upwards by the flowing fluid. In steady flows, all the positions assumed by the float in the tube are characterized by the balance of the three forces acting on the float itself: gravity force, Archimedean force, and drag force. Thanks to the uniformity of the tube taper, the flow rate is proportional to the annular area, which varies according to the position (height) of the float while, consequently, velocity and drag force remain constant. This gives the direct relationship:

$$Q_v = KZ \qquad (5)$$

where K is a proportional constant, and Z is the height of the float. A series of grooves or vanes on the side of the float cause it to rotate (hence, "rotameter") so that it keeps a central position in the tube. The rotameter is a very convenient low cost device of moderate accuracy for medium (or very low) flow rate, operating pressure, and temperature $P_{max} \leq 20$ atm, $T_{max} \leq 200°$ C) both in clean gas and liquid flows. These meters are also characterized by constant low pressure losses, a very useful peculiarity when handling low flows [40]. They require a strictly vertical mounting and have rangeability as wide as 10:1.

A more sophisticated version of this device has a metal tube containing a metal float. A remote electrical device indicates the height of the float and, consequently, the flow rate. This model of rotameter can be used at much higher pressures and temperatures and is therefore more suitable for process control purposes. In these meters the movable member (float) is restrained solely by gravity. In another type of rotameter a spring is used to restrain the movement and thus create exceptionally high rangeability.

The cylinder and piston flowmeter belongs to the variable area class because its operating principle is similar to the rotameter. In fact, both employ a float that is moved by the flow and in both cases the force is constant in value. In the cylinder and piston design, the float exerts a constant downward force, and the position of the piston is due to the difference in pressure between the sides of the float, which works like a piston. If the downstream flow gets heavy, the pressure is reduced on the loaded side of the piston, which is forced up, thereby increasing the number or area of the vertical openings through which the fluid can flow until it reaches balance. The main uses of meters of this type are in measuring high-viscosity fluids such as fuel oils, tar, and chemical liquors, corrosive fluids or materials that might clog lines, and fluids whose flow characteristics are not well known. Although their accuracy is only moderate, their rangeability is wide and their cost is rather low.

A short list of the advantages and limitations of variable area flowmeters is given in Table 3.

Table 3

Variable Area Flowmeters

Advantages	Limitations
1. Low cost	1. Rotameter: must be mounted vertically
2. Low and constant pressure losses	2. Not suitable for high pressures and temperatures
3. Suitable for very low flow rates	3. Rotameter: dirt sediment on glass can make reading difficult
4. Rangeability 10:1	4. Clean fluids only
5. Capable of measuring fluids of varying density and viscosity (compensation given by float design)	5. Cost rises considerably with extras (protection shields, panel mounting, etc.)

Positive Displacement Flowmeters

All positive displacement meters operate by dividing up the fluid into a number of discrete "packets" and then counting the total number of packets that pass through the meter in a known length of time. An output shaft drives through gearing to a local display counter; by selection of suitable gearing, a readout in the required volumetric units can be obtained. A pulse generator, either optical or electromagnetic, also may be fitted for transmission to a remote control room.

Several very different sub-types of these meters exist, which, in turn, because of their limited technological tolerance, require calibration to establish true swept volume.

Some positive displacement meters are extremely accurate, repeatable, and have good rangeability (20:1). They are commonly used for metering the total quantity of fluid flowing rather than the flow rate. The most highly engineered meters are very accurate and therefore are very widely used for metering liquids such as fuel oils and other hydrocarbon products. As an example, all meters in gasoline pumps are of this type. They are usually used with small-size pipe because the initial cost of the meter rises dramatically with larger pipe diameters. The most accurate types are expensive and precision-made, but cheap mass-produced types are also available and are widely used as domestic water meters.

Most positive displacement meters are used with liquid, but several versions are also available for low and high pressure gases [1]. Nevertheless, such meters also present some great disadvantages. For example, the moving parts cannot fit perfectly in one another nor in the stator without producing an unacceptable degree of friction. Thus, the metering fluid must be clean or filtered so as to avoid erosion of the moving parts and/or any sediment that might modify tolerances or the volume of chambers. Their internal construction can take a large number of different forms, but only two types, the rotary vane and the nutating disk, will be described here.

The basic operating principle of the rotary vane meter is simple. In fact it consists of a continuous rotating piston (rotor), driven by the fluid, that in its motion provides a discrete number of isolation chambers delivering a quantity of metering fluid (packets). The isolation chambers are formed by the space among the housing, the rotor, and two adjacent vanes moved radially by a cam. Springs retract the vanes at the end of each cycle. The sum of the revolutions of the rotor are registered in volumetric terms.

In the nutating disk flowmeter the isolation chambers are formed by the conical space between the housing and a circular disk attached to a ball. A vertical partition attached to the disk restrains it from rotation, but allows it to nutate or wobble. The nutating motion is generated when the fluid enters the isolation chamber alternately above and below the disk. Complete separation between inlet and outlet chambers volumes is always achieved by one dedicated disk diameter line. The display counter is activated by a shaft mounted through the center that moves along a circular path.

A short list of the advantages and limitations of positive displacement flowmeters is given in Table 4.

Table 4

Positive Displacement Flowmeters

Advantages	Limitations
1. Good accuracy and rangeability	1. Regular maintenance in service required
2. Very good repeatability	2. High pressure losses
3. Accuracy virtually unaffected by upstream pipe conditions	3. Moving parts subject to wear
4. Suitable for high viscosity fluids	4. Not suitable for dirty, nonlubricating or abrasive liquids
5. Read out directly in volumetric units	5. Expensive, particularly with large diameters

Inferential Flowmeters

The difference between positive displacement type meters and inferential meters is that the latter do not capture a discrete volume of fluid but rather infer the volumetric flow rate from the action of the fluid on a bladed rotor. A turbine meter basically consists of a bladed rotor suspended in a fluid stream with its rotation axis perpendicular (for low and medium flow rates) or coaxial (for high flow rates) to the flow direction. The rotor is driven by the fluid impinging on the light, flat blades, producing an angular velocity (*n*) proportional to the average fluid velocity over a wide range of volumetric flow rates:

$$Q_v = Cn \tag{6}$$

where C is a proportional factor. The linearity of the above relationship produces a fairly wide rangeability of turbine meters, which the various manufacturers quote as 10:1 to 30:1, with high repeatibilities of $\pm 0.1\%$ and common accuracies of $\pm 0.5\%$ of actual flow.

Mechanical or non-contact techniques (such as electromagnetic, RF proximity, or opto-electronic) are typically used to transform the rotation of the turbine rotor into the metering signal suitable for transmission. The use of these pick-off systems and the technological constraints due to the lightness of the rotor, the bearings, and to the high performances required (especially in industrial fluids) cause turbine meters to be essentially high-performance but also high-cost devices. Thus, they are gradually ousting positive displacement meters in some applications (e.g., natural gas), although the latter are still used in process control when high accuracy is required. Turbine flowmeters are used for accurate flow measurements of clean gases and liquids [42-44], though some manufacturers offer devices specifically designed for steam measurements. Two different turbine rotor designs are required for liquids and gases because of the great difference in density and velocity between the two fluid states.

Turbine flowmeters are sensitive to the velocity profile and the presence of swirls at their inlet [44]. The greatest problems with liquid service are caused by the detrimental effects of overspread when the liquid flashes or when slugs of vapor or air enter the line. These events shift meter calibration, especially for liquids containing small amounts of air, and at the same time produce blade wear and bearing friction [45].

Low-velocity performances are affected by velocity profile, tip clearance, friction across the blades, bearing friction, and retarding torques due to hydrodynamic and bearing friction drags. Thus, with sizes below a certain diameter, liquid turbine meters become less and less accurate, so that their use is not recommended at all in very small sizes [3].

With large pipe or ducts, turbine insertion meters can be used successfully. These are simple and economical devices, though they can grant only reasonable accuracy, since, like all insertion-type meters, they infer the volumetric flow rate from a local velocity measurement [34], which is only theoretically correct.

A short list of the advantages and limitations of inferential flowmeters is given in Table 5.

Table 5

Inferential Flowmeters

Advantages	Limitations
1. High accuracy	1. Moving parts subject to wear
2. Rangeability 10:1	2. Can be damaged by overspeeding
3. Very good repeatability	3. High temperature, overspeeding, corrosion, abrasion and pressure transient can shorten bearing life
4. Low pressure drops	4. Rather expensive
5. Versatile and suitable for operation under severe conditions	5. Filtration required in dirty fluids

Fluid-Dynamic Flowmeters

Fluid-dynamic flowmeters are the last of the energy extractive meters to be discussed here. These are the newest in the category (energy extractive), having appeared on the market in the late 1960s [46], yet they can still be defined as "emergent" among the industrial meters. Two different fluid-dynamic phenomena are involved in this type of meter: vortex formation (vortex shedding and vortex precession) and the Coanda effect.

Vortex formation is due to the typical instability of the wake that forms behind an obstacle (bluff body) placed perpendicular to the flowing fluid. Moderate flow will follow the irregular contours, but as velocity increases, the stream separates from the body. Fluid particles in the boundary layer next to the bluff body accelerate, then stop and reverse to form an eddy or vortex. This vortex grows and separates from the surface. Vortexes form on alternate sides of the body, and as they separate and move downstream they form the "Von Karman Vortex Trail". The frequency at which these vortexes are shed (and the related oscillating distribution of fluid pressure and density) is directly proportional to the fluid velocity, thus providing the basis of a vortex-shedding flowmeter.

The Coanda effect is a phenomenon of fluid jet attachment to one of two side walls, together with the fluidic action of the fluid itself. In the fluidic oscillator flowmeters based on this phenomenon, a straight jet cannot keep its course in the presence of two walls but necessarily runs along one of them. This principle is used to transform a bistable fluidic amplifier into a self-oscillating system. A feedback loop, obtained by diverting a small quantity of metering fluid, is devoted to self-induce the oscillation of the flow between the side walls of the meter body. A proper geometrical design of walls, ports, and feedback passage, in which a sensor (usually a thermistor) is heated, causes a digital or pulse output with a frequency proportional to the fluid velocity.

The oscillating behavior of both fluid-dynamic meters described above can be realized through the concept of the complementary relationship between jet and wake [47]. Dimensional analysis and experimental results support this thesis of structural complementarity and the kinematic similarity of the two different types of meters. These same results confirm the linearity of the relationship between frequency and volumetric flow rate in the Reynolds number range where the Strouhal number (St) can be considered as a constant. In the two different meters (vortex-shedding and fluidic oscillator) the Strouhal number is defined by:

$$St = \frac{fh}{V} \tag{7}$$

where f is the frequency of vortex or jet oscillation, h is a characteristic width of the bluff body or of the loop, and V is the flow velocity. Because of this similarity, both flowmeters show the following advantages: frequency output is linearly proportional to the volumetric flow rate; both have wide rangeability; there are no moving parts; and calibration is insensitive to the physical properties of metering fluids.

The metrological features of these two types of meters, though similar, are not exactly the same because of differences in construction and different operating limitations [48, 49].

The following are worth pointing out:

(1) In vortex-shedding meters the minimum Reynolds number required is typically 10^4; in fluidic oscillator meters it is as low as about $3 \cdot 10^3$.

(2) The applicability of fluidic oscillator meters is restricted to clean liquids (solid particles content lower than 2%) because of the obvious limitations due to the possible blockage of feedback loops. Their rangeability is up to 20:1. Vortex-shedding meters, however, can be more widely used in liquids, gases, and vapors. They grant rangeability up to 20:1 in liquids and in gases, even if many authors claim an extended flow range in gases of 100:1 [8, 11, 48]; and display better adaptability to measuring dirty liquids, provided that such liquids have Newtonian behavior.

(3) Vortex-shedding meters have good accuracy, which is constant in the whole range (±0.5 to ±1% rate with liquids; ±1.5 to ±2% rate with gases and vapors). The accuracy of fluidic oscillator meters is not constant in the whole range, with lower values than vortex-shedding types (±1 to ±2% FS).

It is also worth remembering that both types of meters require long straight pipes upstream and downstream and that vortex-shedding meters are most suitable for pipes with fairly large cross section (25 to 200 mm), while fluidic oscillator meters allow use with a maximum diameter of 100 mm.

Some models of insertion-type vortex-shedding meters suitable for large pipes are available.

Vortex-shedding meters are constructed differently by various manufacturers, especially in the designs of the bluff body and the vortex detecting system, which can consist of either pressure sensors (piezoelectric or strain gage detectors) or velocity sensors (heated thermistor or ultrasonic beam detectors). It appears that fluidic oscillator meters at the time of this writing use only thermal sensors, although, in theory and owing to their similar working principles, sensors of other types might also be used.

The vortex precession flowmeter, also called a swirlmeter, is another model of fluid-dynamic flowmeter based on the vortex formation phenomenon. In operation, a set of inlet fixed swirl blades adds a tangential velocity component to the axial velocity of the entering fluid, thus producing a swirling motion. Flowing through a Venturi-like passage, this swirl becomes like a vortex filament that, as result of a secondary rotation of vortex core generated by pressure distribution and body shape, follows a helical path. At a fixed downstream station, a sensor (usually a thermistor) detects the pressure or the velocity fluctuations due to the helical path. These fluctuations constitute the frequency of the resulting precession-like motion [50, 51], which is linearly proportional to the flow rate.

The swirlmeter's accuracy and rangeability are comparable to those of the vortex-shedding device. However, the vortex precession flowmeter is obsolete, usually available only for gas flow application, just in one size.

Nevertheless, some technical literature boasts its suitability also for clean (Newtonian) liquids. One major disadvantage of the swirlmeter is that pressure losses are approximately five times as much as in vortex-shedding flowmeters. This, as well as some other construction problems, has limited the development and distribution of these meters much to the advantage of the simpler vortex-shedding meter, which is the most widely used of the three fluid-dynamic types.

A short list of the advantages and limitations of fluid-dynamic flowmeters is given in Table 6.

Table 6

Fluid-Dynamic Flowmeters

Advantages	Limitations
1. Good accuracy	1. Not suitable for dirty or abrasive fluids
2. Usually wide flow range	2. Straight upstream pipe required equal to 10 times pipe diameter or longer
3. Used with liquids, gases, slurries (except for the swirlmeter, which is suitable essentially for gases)	3. Limited size range
4. Minimal maintenance (no moving parts)	4. Limited by lower velocity ($Re > 10^4$)
5. Good linearity over the useful range	5. Appreciable pressure drops (especially for swirlmeter)

Tracer Flowmeters

Tracer or tagging techniques can be defined as special methods used to infer flow rate from the transit time of the tagging signal, which is why they are discussed here, following the energy extractive types and before the energy additive approach to flow measurement. In fact, tagging techniques can be either energy extractive or energy additive, according to the type of marker they use.

Tracer meters are based on the principle of adding a "marker" or "tracer" to the fluid at one point or section of the flow path, then detecting its presence at a second point at a known distance downstream. The transit time provides information on the flow rate.

Tracer or tagging designs can be grouped into two general categories: (a) the marker is introduced into the flow by injection or induction, and (b) the marker exists in the flow in the form of natural dust or dirt or natural turbulence (vortices or eddies). The devices that use already existing tags or markers are: (a) Laser Doppler Anemometers (LDA), which use dust or dirt as natural tracers and are more frequently used in laboratory research than in process control [52]; and (b) cross-correlation flowmeters, which use turbulent eddies or clumps of particles moving along with the velocity of the flow. Sensors now being used in cross-correlation flowmeters are either optical or ultrasonic; their principle of operation is based on the maximization of the cross-correlation function $R_{xy}(\tau)$ between the positions of the two sensors, whose distance is l:

$$R_{xy}(\tau) = \int x(t - \tau)\, y(t)\, dt \tag{8}$$

where τ is the cross-correlation time lag, and $x(t)$ and $y(t)$ are the signals that each sensor gives when τ is equal to the transit time of the tracer [53]. Hence, the flow velocity, V, is given by:

$$V = \frac{l}{\tau} \tag{9}$$

In the past, cross-correlation flowmeters were limited by the exceedingly high costs of their computing systems, but the use of microprocessors is allowing them to develop to a stage where they appear suitable for solving many industrial and environmental flow measurement problems where noncontamination is required.

As for the techniques where a marker must be inserted, the difference between injection and induction is that injection requires an entry mechanism through the pipe wall to the fluid, whereas induction can operate without it. Typical injection methods use the following as markers: salt solutions, hot or cold fluids, gases, radioactive isotopes, and either fluorescent or opaque dyes.

In induction methods the markers are either ionized gas or magnetized nuclei. The best known induction method is the Nuclear Magnetic Resonance (NMR) flowmeter; it uses a tagging technique by which a nuclear magnetization is induced in some fluid nuclei having a magnetic moment. These nuclei become the markers and are detected subsequently between two stations that are a known distance apart. This technique is available for fluids containing hydrogen or fluorine compounds with a nuclear magnetic moment. In operation, fluid first enters a magnetizer section where some degree of nuclear magnetism is imposed. At the second station the fluid enters a second magnetic field where resonance is obtained by applying a radio frequency signal. A modulating magnetic field creates demagnetized pockets, and the modulating effects are detected by a final RF detector coil. The modulation and detected signals are fed into a phase comparator that adjusts the modulation frequency so as to maintain a fixed phase difference between modulation signal and detected signal [54]. The modulation frequency is then a measure of flow rate:

$$Q_V = kf \tag{10}$$

where k is a proportional constant and f is the modulation frequency.

Using the nuclear magnetization imparted to the fluid, the flow measurement is made from outside the pipe wall, and no electrical, mechanical, or optical contact with the fluid is required. This results in a straight pipe, nothing-in-the-stream design in a flowmeter with the appearance of a thick-walled pipe spool. NMR flowmeters are independent of such fluid properties as conductivity, temperature, viscosity, density, and gas or solid entrainment.

A short list of the advantages and limitations of tracer flowmeters is given in Table 7.

Table 7

Tracer Flowmeters

Advantages	Limitations
1. No moving parts	1. External devices or electrical power is required
2. Independent of the fluid properties (temperature, density, viscosity, etc.)	2. Some models more suitable for detecting velocity than for metering flow rate
3. Good performance devices	3. Possible noises and dispersions
4. Independent of flow regimes (except in the region of the laminar/turbulent boundary)	4. Response time of 2 to 3 seconds
5. No pressure losses	5. Relatively high cost

Electromagnetic Flowmeters

According to the latest trends of the market, meters in the energy additive category are where most new applications and developments are likely to take place, both in the number and types of models. They are dealt with from this section on.

Electromagnetic flowmeters (emfm) are based on Faraday's Law and are among the earliest and simplest powered measurement systems. Faraday's Law states "the voltage induced across any conductor as it moves at right angles through a magnetic field is proportional to the velocity of that conductor":

$$e = V \times B \tag{11}$$

which in scalar terms is

$$e = kVDB \tag{12}$$

where:

e = the emf induced in the fluid (liquid)
k = a constant
V = the fluid velocity
B = the magnetic flux density
D = the distance between electrodes, i.e., the pipe diameter

In accordance with this law, the essential scheme of an emfm consists of an electrically conducting liquid (with electrical conductivity greater than 5 μS/cm) flowing in a tube made of nonmagnetic material. The liquid flows between two electrodes that are in contact with the liquid and located perpendicularly to the flow and to the lines of the magnetic field (produced across the tube by exciting coils placed outside the tube). The magnetic field, in turn, is perpendicular to the flow direction. The motion of electrolytes down the tube through the magnetic field produces a potential difference between the electrodes; this induced voltage output is proportional to the liquid velocity and gives an indication of the flow rate. In reality, each electrolyte contributes to the potential difference emf to a degree that depends on its own velocity and position in the magnetic field. Therefore, if the magnetic field is constant and symmetric and the distance between the electrodes is fixed, the Faraday-induced voltage is directly proportional to the average velocity of this array of "batteries" or generators" (uniformly distributed in the liquid only if the velocity profile is axisymmetric), giving an output linearly related to the volumetric flow rate [55].

To sum up, a magnetic flowmeter possesses the following characteristics: (a) it is unaffected by the variations in the thermophysical properties of the metering liquids, such as density and viscosity; (b) it displays a very wide flow range, the measurement being independent of Reynolds number; (c) it is obstructionless so no additional pressure losses are imposed by the meter; (d) it is suitable for measuring very dirty liquids.

Furthermore, emfm are available in special versions suitable for metering extremely corrosive or aggressive liquids. In such meters the tube is made of abrasion-resistant rubber, polyurethane, or Teflon.®

The electrodes are of stainless steel in standard models, but types in more exotic materials such as tantalum or platinum iridium are also available for smaller sizes. They are subject to fouling if dirty liquid is being metered, and therefore ultrasonic cleaning devices must be incorporated to overcome this problem.

Conventional emfm have accuracy expressed as a function of full scale, typically 0.5 to 1% FS. Nevertheless, the simplicity of emfm construction schemes is only apparent. In fact, it is necessary to take into account the problem of spurious inter-electrode potentials and the related problem of transformer signal, as well as the difficulty in predicting the sensitivity of the meter. The presence of spurious signals can be due to: (a) cell-like action if the chemical compositions of the electrode contacting surfaces are not equal; (b) thermoelectrical effects if the temperatures of the two electrodes are not the same, and (c) polarization and/or electromotive effects on the electrode surfaces. All these spurious types of voltage are unpredictable and can vary in time, but their effects can be overcome through ac excitation of the magnet coils. Sinusoidal, square wave, pulse, and trapezoidal are among the most widely used types of ac excitation [56].

The sensitivity of an industrial emfm can be predicted only with a random variation of 10%. This sensitivity variation is due to the geometry, to the electrical properties of the tube and/or magnet core, and to the current coil supply variations, so that each meter must be calibrated individually. Such problems contribute to the relatively high cost of such meters. Thus, whenever feasible, more economical meters are usually preferred.

In addition to the unsuitability of emfm for gases and nonconductive liquids, further limitations to their use are due to temperature limits and to the effects of velocity profile. Maximum temperatures range up to 200°C depending on the materials of construction selected and the related cooling problems. The effects of velocity profile are related with the extreme sensitivity of the meter to asymmetrical flow profiles. This drawback can be overcome by mounting the flowmeter with upstream and downstream straight pipes in pure liquids and by making sure that the pipe is completely filled with liquid with no air or gas bubbles present. In addition to these cautions, vertical mounting is advisable in dirty liquids or slurries to avoid stratification.

More expensive models are available that allow use with nonsymmetrical velocity profiles by means of an adequate shape of the magnetic field.

Some versions of emfm insertion-type meters are suitable for very large size pipes.

A short list of the advantages and limitations of electromagnetic flowmeters is given in Table 8.

Table 8

Electromagnetic Flowmeters

Advantages	Limitations
1. Obstructionless flow	1. Liquid must be electrically conductive
2. Unaffected by viscosity, pressure, temperature, density	2. Not suitable for gases
3. Good accuracy and wide rangeability (30:1)	3. Expensive, particularly in small sizes
4. No Re constraints	4. Can be sensitive to asymmetric flow profile
5. Suitable for slurries and corrosive, nonlubricating, or abrasive liquids	5. Calibration is required

Ultrasonic Flowmeters

This second group of powered flowmeters is the most promising in the field of nonintrusive (i.e., with no mechanical obstruction) meters. Nonintrusive meters are energy savers and suitable for use with any type of liquid, regardless of the variations of its thermophysical properties, though some ultrasonics are affected by Re.

A further advantage such meters may offer is that they can be constructed in clamp-on types whose installation is obviously simple. Therefore, the larger the pipe size, the more economical the clamp-on measurement device.

The sonic flowmeter class is subdivided into four basic types: time of flight, Doppler, cross-correlation, and swept-beam meters. The most common types used in process control are time-of-flight (TOF) and Doppler ultrasonic flowmeters. In both types a piezoelectric crystal is excited by electrical energy at its mechanical resonance, thus emitting a sound wave, which, travelling at the speed of sound in the medium, is used to infer the flow rate. The crystal is placed in contact with the fluid (wetted or inserted transducers) or else mounted outside the piping (clamp-on transducers) [57].

In the time-of-flight type, a transmitter beams a high frequency (approximately 1 MHz) pressure wave (ultrasonic wave) so as to form a fixed cross angle with the pipe axis. The transit time employed by the wave to reach a receiver placed on the opposite pipe wall depends on both the velocity of sound in the fluid and whether the wave is moving with or against the flow. Flow rate information is obtained from the measured time:

$$t = \frac{L}{c_0 + V \cos \phi} \tag{13}$$

where:

t = the wave propagation time
L = the distance between transmitter and receiver

c_0 = the sound velocity in fluid (in water $c_0 = 1481$ m/s, when t = 20°C

ϕ = the angle between flow direction and beam direction

Thus, velocity (and consequently flow rate) is given by the relationship:

$$V = \frac{1}{\cos \phi} \left(\frac{L}{t} - c_0 \right) \quad (14)$$

Unfortunately, in the process field, the speed of sound in a fluid not only is usually unknown, but also can vary with fluid state properties such as temperature and density. In order to avoid this effect, two series of sonic pulses with known travel frequency are employed. The upstream series is subtracted from a similar downstream series of sonic pulses. The measured difference

$$\Delta f = \frac{2V \cos \phi}{L} \quad (15)$$

between the frequencies is a direct function of the flow velocity, which is independent of the velocity of sound:

$$V = \frac{L \, \Delta f}{2 \cos \phi} \quad (16)$$

Therefore, the volumetric flow rate is proportional to the difference between frequencies:

$$Q_v = k \, \Delta f \quad (17)$$

where k is a constant.

A wide variety of time-of-flight ultrasonic flowmeters are available, but the most remarkable differences usually lie in the number of beam paths traveling across the pipe [58]. A single beam averages profiles along the beam and not across the pipe area. This makes the single-path measurement strongly dependent on the Reynolds number, i.e., on the velocity profile.

Multipath devices, in averaging along several paths, reduce velocity profile dependency but tend to become very expensive and are normally used only in very large pipes.

In "wetted" flowmeters, after ascertaining the compatibility between sensors and possible corrosive or coating properties of the fluid, it is crucial to locate the sensors properly so as to gain high sensitivity to swirl or eddy formation on the part of single-path and multipath flowmeters. In the clamp-on version, once the perfect transducer alignment has been determined, it is very important to discover the correction function in terms of Reynolds number and to place the transducers in the right position, which is far away from any source of fluid-dynamic disturbance [59].

Time-of-flight flowmeters are generally used in clean liquid applications, where the ultrasonic beam is not attenuated or continually interrupted by fluid particles. Some manufacturers claim success with rather dirty fluids, though sufficient experimental data are not available to confirm such performances. Claimed accuracies range from ±1% to ±4%, depending on design and on the right utilization of the meter.

The second type of ultrasonic flowmeter is based on the well-known Doppler effect. When a wave beam travels into a nonhomogeneous fluid, some energy is scattered back by solid particles or air bubbles entrained in the flow. The relative motion of these discontinuities produces a frequency shift of the scattered wave, which is received and analyzed by a transducer. This different frequency, known as Doppler-shift, Δf, is linearly proportional to the fluid velocity, i.e., to the flow rate:

$$Q_v = C \, \Delta f$$

where C is a constant.

This phenomenon (Doppler effect) is employed to measure the fluid flow whether in the Doppler ultrasonic flowmeter utilizing an acoustic beam, or in the Laser Doppler Anemometers (LDA) utilizing a laser light beam. The Doppler flowmeters are very widespread in the clamp-on configuration, which has the great advantage of being mounted in line without disturbing and stopping the process. Moreover, in this type of ultrasonic flowmeter the transmission angle of the sonic beam is not as crucial as it is for TOF flowmeters. This feature is emphasized also by the cheaper versions of Doppler meters that allow for just one transducer, which acts as both transmitter and receiver. In a clamp-on Doppler flowmeter, the transducer transmits a continuous-wave ultrasonic beam through the pipe wall and into the fluid.

In operation Doppler-type ultrasonic flowmeters require a non-porous pipe material through which flows a sonically conductive liquid containing a quantity of solid particles of air bubbles sufficient to scatter signals toward the sensor. The minimum quantity of these sonic discontinuities may vary from 0.005% to 0.1% per volume, depending on the fluid and on the size of the entrained particles. Because of the velocity profile, accuracy depends on particle concentration and distribution. If uniformly distributed, such particles influence the depth of penetration of the acoustic beam. For instance, if there is a high concentration of particles, the flowmeter will read only the slow-moving reflectors near the wall while undervaluing the flow rate; the same occurs in very low concentrations where undervaluing is due to the further contribution of the slow-moving reflectors on the opposite wall of the pipe. Such drawbacks are partially overcome by using two transducers on the opposite walls of the pipe [60]. Accuracy is also influenced by the difference in velocity of the particles and the fluid and by the pipe size diameter where, in accordance with what was stated previously, the measured velocity will not coincide with the average fluid velocity. For all these reasons, industrial Doppler flowmeters display low accuracy (±2.5 to 5% rate), even if they are characterized by great repeatibility (±1% FS) in a given situation. Currently, ultrasonic flowmeters have only limited application in the field of gas flows; they are used primarily with liquids, where they are gaining popularity. It goes without saying that all ultrasonic devices require a full pipe flow for correct operation.

A short list of the advantages and limitations of ultrasonic flowmeters is given in Table 9.

Table 9

Ultrasonic Flowmeters

Advantages	Limitations
1. Non-intrusive, obstructionless	1. Maximum temperature 150°C
2. Wide rangeability	2. Particular fluid conditions are required (TOF-type: clean liquids; Doppler-type: particles or impurities in the stream)
3. Easy to install (especially for the clamp-on version)	3. Not very high accuracy (about ±2% FS)
4. Cost virtually independent of pipe size	4. Doppler flowmeter clamp-on type requires a pipe of homogeneous material (cement, cast iron, fiberglass must be avoided)
5. The flow measurement is bidirectional	5. Periodic recalibration is required

Mass Flowmeters

As the cost to produce chemicals, petrochemicals, and petroleum products rises, the need for more accurate measurement becomes increasingly more important in order to optimize processes and minimize waste. Since rational usage of energy is vital, mass flow measurement has been found to be the most effective way to optimize consumption. Two main techniques are used for measuring mass flow rate: the direct method (mass flowmeter) and the indirect method (flow computer). The flow computer infers mass flow rate by processing the data supplied by a certain number of instruments (a volumetric flowmeter and pressure and temperature sensors employed to obtain thermophysical properties such as density). Therefore, the indirect method is the less accurate because it entrains the measurement uncertainties of each device [61]. Direct mass flow measurement is the more accurate not only because just one instrument is employed, but also because of its intrinsic characteristics. In fact, mass flowmeters are not affected by thermophysical parameters.

"Not True" Mass Flowmeters (Thermal Flowmeters)

This group of flowmeters belongs to the wider emergent category of meters that measure the mass flow rate, which, in many industrial and commercial applications of flow measurement, is clearly much more relevant than volumetric flow rate. Thermal mass flowmeters, otherwise called calorimetric flowmeters, infer mass flow rate from the mass and the energy balance equations of the fluid. These balances depend on the thermal behavior and properties of the fluid; therefore, the measurement can be done only if these properties (such as thermal conductivity and specific heat) are constant and well-known and if the metering fluid is the same as the one used for calibration (otherwise, it might be possible to relate the metering fluid to the calibration fluid, though different). This is the reason for classifying such meters as "not true" mass flowmeters [3].

Thermal mass flowmeters work on the principle of heat transfer by the fluid flow. They consist of three elements arranged consecutively along the direction of motion: an electrical heater is placed between highly accurate temperature sensors installed, respectively, upstream and downstream of the heater. If the thermal properties of the fluid being metered are constant and known, the difference between the two temperature readings is proportional to the mass flow rate [62]. If heat absorption on the part of the pipe walls is negligible, the fluid energy balance is expressed by following relationship:

$$\dot{Q} = \dot{m}\, c_p(t_a - t_b) = RI^2 \qquad (19)$$

where:

$\dot{Q}$ = the thermal power
$\dot{m}$ = the mass flow rate
c_p = the specific heat at constant pressure
t_a = the temperature of the fluid after the heater
t_b = the temperature of the fluid before the heater
R = the resistance of the heater
I = the electrical current

Thus, the temperature rise in a stream gives a measure of mass flow rate. In theory, correct operation requires a constant $\rho \cdot c_p$ product (density × specific heat) as well as efficient mass transport (i.e., ideal heat transfer), allowing a perfectly uniform temperature profile at the downstream measurement station.

The thermal mass flowmeter devices currently available can be divided into three different types. The first configuration has the temperature sensors and heater mounted outside the pipe so that they never come in contact with the fluid. In these meters only the boundary layer of flow near the pipe wall is heated. The temperatures upstream and downstream of the heater are also detected by external thermometers. Although the flow rate in the boundary layer is not the same as in the main flow, this measurement may be used to infer the main flow rate by means of the proportional constant. The dynamic response of this approach is not satisfactory because of the great thermal inertia of the walls. These types of meters are useful in small pipe sizes but cannot be used with large fluid flows because the energy consumption would be prohibitive. For this application, meters have been recently designed without a heater but with two temperature sensors mounted on the outside diameter of the tube, which never come into contact with the fluid. An upstream sensor detects the temperature of the flowing stream, while a downstream sensor is kept at a fixed constant temperature a certain number of degrees above the upstream sensor. This type of meter is used in liquid, gas, and slurry service with a claimed accuracy of ±1% FS.

For a very low mass flow rate of clean liquids and gases based on the same principle of operation, a second type of thermal mass flowmeter is available, displaying an externally heated bypass capillary tube. Two thermocouples installed outside the capillary tube sense the variation of temperature distribution along the tube itself. In zero flow conditions this distribution is symmetrical, but it becomes asymmetrical with flow. The bypass is realized across a laminar element placed in the main flow tube in which the metering fluid passes. The laminarity of flow guarantees the proportionality between the two split flows. The accuracy of these devices is comparable to that displayed by the previous version (±1% FS).

A third type of thermal mass flowmeter, the "insertion-type meter," has recently been very successful for gas applications in large ducts. These devices basically consist of two small platinum resistance temperature detectors (RTDs) positioned in the sensor probe beside a heating element that heats only one of the RTDs. The temperature difference between the RTDs is greatest at zero flow and decreases as the flowing gas passes across the sensing element, cooling the heated RTDs. The heat dissipation is a logarithmic function of the mass flow rate if the gas properties are constant. The validity of these insertion meters is due to the experimental fact that, at the downstream end of a long length of straight pipe, the mean velocity of the flow is approximately the same as the velocity at a point one quarter of a radius from the pipe wall. Consequently, it is possible to economically measure the flow in very large pipes by inserting a small meter at the quarter-radius position (or "three-quarter radius" position, as it is sometimes termed, depending on whether you view it from the center of the pipe or from the wall). If the insertion meter is a velocity meter (such as a thermal anemometer, a Pitot tube, or any flowmeter insertion version: vortex, turbine, target, electromagnetic, etc.) it infers the volumetric flow rate; if it is an insertion mass flowmeter, by measuring the local mass flow it can infer the total mass flow rate. Insertion thermal mass flowmeters are inherently less accurate than full-bore flowmeters but, whenever a moderate accuracy is acceptable, they can provide a very economical mass flow measuring of gases in large pipes or ducts with great rangeability.

A short list of the advantages and limitations of thermal mass flowmeters is given in Table 10a.

Table 10a

Thermal Mass Flowmeters

Advantages	Limitations
1. No moving parts	1. Meter sensitivity to fluid heat conductivity, viscosity, and specific heat
2. Suitable for large size pipe (insertion type)	2. Energy consumption required
3. Good rangeability (10:1)	3. Mostly gas service (only rarely liquid service)
4. Accuracy: ± 1–2% FS	4. Specific heat of the fluid must be known and constant, i.e., the gas must have a constant composition
5. Low permanent pressure losses	5. Proper operation requires no heat losses due to conductive exchanges through the pipe walls

True Mass Flowmeters (Angular Momentum)

The angular momentum type of mass flowmeter, also called "axial flow mass flowmeter," is one of the oldest and most popular approaches to true mass flow measurement [63].

In operation, a motor-driven impeller, rotating at constant speed, imparts a rotary motion to the flowing fluid, which then passes through a geometrically similar turbine mounted in sequence in the pipeline and is prevented from rotating by an elastic constraint. Both the coaxial impeller and turbine present several straight passages parallel to the axis of the pipe realized in an annular peripherical space. In service, the fluid passing through the impeller acquires a constant angular momentum in addition to its axial velocity. The turbine, or rather its linkage (springs), absorbs this angular momentum, and its angular deflection (or torque) will be a measure of mass flow rate. In fact, the relationship between the mass flow rate (m), the torque (T), and the constant angular velocity (ω) is:

$$\dot{m} = \frac{T}{k\omega} \quad (20)$$

where k is a constant.

From relationship (20) it results that the mass flow rate can also be inferred by keeping the torque constant and varying the angular velocity, which becomes a measure of mass flow. Hence, another type of angular momentum device utilizes a feedback control to vary and measure the angular velocity of the impeller.

A further variation can be obtained measuring the mass flow by detecting only the power required to move the impeller. An angular momentum mass flowmeter based upon the above said principle does not need the turbine and has only a particular impeller, which is like a rotor of an electric motor. The relative stator is winded on the outside of the pipe. The relationship between the electrical power, P, and the mass flow is:

$$P = m\,\omega^2 k \quad (21)$$

where ω is the angular velocity of the rotor and k is a constant. The power transferred to the fluid by the impeller rotating at constant velocity is directly proportional to mass flow.

Angular momentum mass flowmeters generally possess the same potential accuracy of all true mass flowmeters (about ±0.5%). They are usually applicable to liquid and gaseous flow. The greatest limitations of this type of meter compared to the other static-type true mass flowmeters lie in their rotating parts, which require maintenance, and in their not being recommended for dirty fluids.

A short list of advantages and limitations of angular momentum mass flowmeters is given in Table 10b.

Table 10b

Angular Momentum Mass Flowmeters

Advantages	Limitations
1. High accuracy: ± 0.5%	1. High maintenance of moving parts required
2. Relatively low pressure drops	2. One-way measurement
3. Suitable for liquid and gas flow	3. Generally used only with noncorrosive fluids
4. Easy to install	4. Difficult choice of adequate spring materials
5. Fairly wide (10:1) flow range	5. High fluid temperatures affect the elasticity of the system

Pressure Differential-Type Mass Flowmeters

In principle, pressure differential-type mass flowmeters, the hydraulic equivalents of the electrical Wheatstone bridge, are obtained by arranging four identical orifice plates in a hydraulic network. Therefore, they are also called "fluidic or hydraulic Wheatstone bridge flowmeters" [64]. A positive displacement pump, placed on a diagonal and running at a constant speed, recirculates a constant volumetric flow rate from one side of the bridge into another. The pressure rises across the upper (U) and lower (L) orifice legs of the hydraulic bridge are:

$$\Delta p_U \propto C\,\rho\,Q_U^2 \quad (22)$$

$$\Delta p_L \propto C\,\rho\,Q_L^2 \quad (23)$$

where C is the orifice discharge coefficient, ρ is the fluid density, and Q is the volumetric flow rate. Two different operating situations can be met with:

(1) The pump operates a constant volume of flow, q, with q less than the total flow rate Q. The volumetric flow rates in the upper and lower legs are:

$$Q_U = \frac{Q}{2} - q \quad (24)$$

$$Q_L = \frac{Q}{2} + q \quad (25)$$

Since the pressure rise in the pump Δp_p is equal to the pressure difference in the legs, it follows that

$$\Delta p_p = \Delta p_L - \Delta p_U \propto C\rho\,(Q_L^2 - Q_U^2)$$

$$\propto C\rho\left(\frac{Q}{2} + q\right)^2 - C\rho\left(\frac{Q}{2} - q\right)^2$$

$$\Delta p_p \propto 2\,C\rho Q q \quad (26)$$

As mass flow $\dot{m} = \rho\,Q$, Equation (26) becomes:

$$\dot{m} \propto \frac{\Delta p_p}{2Cq} \quad (27)$$

Thus, if the dischange coefficients of the orifices are the same, it is possible to measure mass flow rate as a linear function of the pressure rise Δp_p across the pump.

(2) The recirculating flow rate q is greater than Q; with similar considerations as in (1), it is possible to measure mass flow rate as a function of pressure drop across the bridge. In other words, the differential pressure across the flowmeter system is proportional to the mass flow rate (with $q > Q$).

These types of mass flowmeters are limited in application only to clean liquids with a viscosity less than 50 cP and no Reynolds number constraints, i.e., fairly wide rangeability (50:1). Their accuracy in the measurement of mass flow is very high (±0.5%), which is typical of all true mass flowmeters.

The pressure differential-type flowmeters are suitable for measuring low and very low liquid flows, and they are unaffected by viscosity variations. Thanks to these characteristics, they have been applied in the control of fuel engine consumption.

A short list of advantages and limitations of pressure differential-type mass flowmeters is given in Table 10c.

Table 10c

Pressure Differential-Type Mass Flowmeters

Advantages	Limitations
1. True mass flow measured	1. High pressure drops
2. Linear output signal (differential pressure) proportional to the flow	2. Periodically, orifice must be recalibrated or changed
3. High accuracy (±0.5% of reading) and repeatability (±0.25% of reading)	3. Electrical power is required to move thc recirculating pump
4. Wide flow range, 20:1	4. Not easy to install in process pipeline (bulky meter)
5. Unaffected by viscosity, density, or temperature changes	5. High purchase cost

Coriolis Acceleration Mass Flowmeters

The first commercially available Coriolis acceleration mass flowmeter (1978) employs an obstructionless U-shaped tube and a T-shaped leaf spring as opposite legs of a turning fork [65, 66]. An external electromagnetic driver excites the U-shaped, cantilever-like structure in flexural vibration at its natural frequency, thereby producing a Coriolis acceleration a_C on the fluid flowing in the tube:

$$a_C = 2\omega \times V \tag{28}$$

where ω is the pulse angular velocity of rotation around the axis passing through the inlet and outlet ends of the U-tube, and V is the average velocity of the fluid flowing through the U-tube. The fluid particles in one leg of the tube are subject to the Coriolis acceleration (a_C) perpendicular to the (ω, V) plane. The resulting Coriolis force of inertia, F_C, also goes in a direction opposite to a_C:

$$F_C = 2m\omega \times V \tag{29}$$

where m is the fluid mass. The Coriolis force F_C acts alternately (perpendicularly to the flow path) in opposite directions because of the opposite directions of the fluid velocity V in the two legs of the U-tube, thus causing an oscillating momentum, M_C, around the symmetry axis of the vibrating U-tube. The resulting momentum, M_C, acting around this axis of symmetry produces a twist-type motion, where the deflection angle θ is directly proportional to the mass flow rate for a constant angular velocity, as shown in the following relationship:

$$\dot{m} = \frac{k_s \theta}{4aL} \tag{30}$$

where:

- θ = the deflection angle
- k_s = the angular spring constant of the system
- a = the cross distance from central axis of symmetry of the U-tube to each leg of the tube itself
- L = the longitudinal length of the U-shaped tube

The angular deflection θ of the U-tube is measured by electromagnetic transducers placed in proximity of the band of the leg. The output of these detectors is a difference of phase, i.e., a difference of time (Δt). The mass flow rate, then, should be expressed by a function of Δt:

$$\dot{m} = \frac{k_s}{8a^2} \Delta t \tag{31}$$

where Δt is the time lag between transducers signals. Thus, the mass flow rate is a function of pipe geometry, spring constant (k_s), and the time lag between pulses. To obtain $\dot{m}$, only Δt needs measuring, since the other factors are constant and known and are, therefore, scale factors.

Another type of U-shaped tube Coriolis mass flowmeter, similar to the type described above, has a counterbalance tube in place of the T-shaped leaf spring.

The obstructionless sensor enables the Coriolis acceleration mass flowmeter to measure liquids with varying physical properties such as density, viscosity, and composition. Fluids containing solid particles and entrained vapor or bubbles do not affect the meter performance. Corrosive, abrasive, and very viscous materials are also easily measured. These flowmeters have also been used successfully to measure heavy gas and vapors (i.e., at relatively high pressure), though the main use remains with liquids. As well as in any other type of true mass flowmeter, the accuracy of these flowmeters is very high (±0.5%). Nevertheless, the greatest drawbacks of such early devices consisted in very high pressure losses, which were due to the small diameter tubes that had to be used in order to allow for elastic deformation of the resonant sensor [19].

Such disadvantages were the reason for the appearance of a new generation of Coriolis acceleration mass flowmeters, which consist of twin and parallel sensor tubes oscillating in counter phase. The use of twin pipes allows, on the one hand, the measuring flow rate to be split in two amounts (thus reducing the pressure losses), and, on the other hand, the inertia forces acting upon the system to be balanced. Therefore, these meters are more suitable for use in process plants. In fact, during recent years several types of meters have appeared, all of which make use of twin tubes with various shapes of resonant sensor tube (double U, two oval helically looped tubes [67], two S-shaped tubes, etc.). These meters show lower pressure losses and better metrological performances than single-tube types and grant accuracy ranging from ±0.40% to as high as ±0.15%, depending on the model. Furthermore, pressure losses decrease dramatically in some models where the peculiar construction features allow for the use of large-size tubes. In reality, all these Coriolis acceleration flowmeters are not completely unaffected by changes in temperature of the fluid; temperature variations influence the spring constant, for instance. Therefore, all second generation meters are provided with electronic thermo-compensation.

Some manufacturers and research institutes are engaged in the design and development of new types of Coriolis acceleration mass flowmeters using straight sensor tubes, which is the ideal solution to minimize pressure losses [68, 69].

The main limitation to the use of Coriolis acceleration mass flowmeters perhaps lies in their high cost, which, in spite of high performance, may not be worth the investment and/or long payback period.

A short list of advantages and limitations of Coriolis acceleration mass flowmeters is given in Table 10d.

Table 10d

Coriolis Acceleration Mass Flowmeters

Advantages	Limitations
1. Measures true mass of all liquids, slurries, and foams	1. High pressure losses due to very small size of the vibrating tube
2. Non-intrusive method	2. Very expensive in all sizes
3. High accuracy performance: ±0.5%	3. The system spring constant influenced by the temperature; therefore, compensation required
4. Rangeability 30:1	4. Unsuitable for gases (unless at very high pressures)
5. Bi-directional	5. Calibration required when liquids with density very different from calibration fluid are used

Part 3
Flowmeter Selection Procedure

The main considerations that govern pipeline flowmeter selection are the following: (a) the fluid's nature and its operating conditions; (b) the interaction between the flowmeter and its installation; (c) the long-term performance of each flowmeter and its reliability; and (d) the evaluation of the total flow measurement costs.

The right flowmeter is the one that will do the job adequately at the lowest possible cost. The wide variety of metrological constraints and plant conditions that characterize instrument performance, as well as the great number of pipeline flowmeters in today's marketplace, make flowmeter choice very difficult. If, on the one hand, it is easy to dispose of one or more "right" flowmeters, on the other hand it is certain that many flowmeters are "wrong" for some particular jobs [25]. Perhaps a theoretically "right" flowmeter is unduly expensive, or a flowmeter becomes "wrong" because of unexpected deterioration through time. Therefore, the choice of the most suitable meter must start from a deep knowledge of the operating principles. Such competence enables the instrumentation specialist to draw up a list of "probable" flowmeters, chosen on the basis of the metrological and plant requirements [26]. Moreover, it is up to the specialist to respect the nature of the meter itself, regardless of any distorting advertising campaign [1]. However, it must be pointed out that all too often the data available on flowmeter behavior are not exhaustive, therefore not allowing the specialist to draw up a comprehensive list.

Starting from the list of "probable" flowmeters, a further selection can be carried out (as shown in the flow chart in Figure 2), leading to a shorter list of "feasible" flowmeters. Finally, the meters can be arranged in ranks on the basis of the ratio of total costs to performance. Nevertheless, the top flowmeter in the classification may not be the right one. The instrument engineer must also take into account other factors, such as the vendor-manufacturer's statements on flowmeter service and the company policy on instrumentation [5, 27].

A typical flowmeter selection flow chart is shown in Figure 2, where the various steps leading to the right choice are presented in thick squares. Thin squares show the situations in which each step may occur. The passage from each step to the next takes place after referring to Tables 1 and 11, as indicated in the circles. Table 1 is a short survey of the main characteristics of flowmeters; Table 11 is a explanatory list of the authors' judgments of the suitability or unsuitability of the meters in the most common applications. The labels "suitable," "sometimes suitable," and "unsuitable" are respectively marked by the following keys: ○, ◑, ●.

The symbol ○ means generally suitable or very useful in most circumstances ($\geq 70\%$); ◑ means worth considering in a limited range of cases (30% to 70%); ● means not normally applicable. As shown in Figure 2, the first step in the determination of "probable" meters is related to the nature of the fluid in the case. Fluids may exist in liquid, gas, or vapor phases, and the phase or combination of phases strongly influences the measurement method. Most types of flowmeters, in fact, were designed to be used with pure single-phase fluids, particularly water-type liquids. Therefore, only few meters (see Table 1) can be used in measuring other types of fluids (gases or vapors), and in this case the initial list of "probable" flowmeters is rather short. For instance, magnetic flowmeters are not applicable in gas and vapor service since they operate only with liquids having electrical conductivity greater than 5 μS/cm. Similarly, some other types of flowmeters, such as the ultrasonic and the Coriolis acceleration types, though "right" in theory are very seldom used because they may require high pressure gaseous fluids.

Multi-phase fluids can be homogeneous or heterogeneous. In the first case (for example, when a condensate accompanies a liquid), most meters are suitable only when the concentration of the secondary phase (the unrequired one) is low or at least negligible, and, therefore, the stream can be treated as single-phase. On the contrary, if the concentration reaches high or significant values, the majority of intrusive meters become unsuitable because such values would affect the working and the reliability of the meter itself.

In the case of heterogeneous multi-phase fluids (for example, when solids are entrained into the liquids), the instrumentation specialist faces many problems. For instance, when the concentration of solid particles or of gas bubbles is high and steady, the secondary phase

	requirement / type of flowmeter	O	V	N	E	P	AP	TR	L	R	CP	RV	ND	TU	VXS	VXP	FO	NMR	EM	TOF	DF	TH	WB	AM	CA
THERMO-PHYSICAL FEATURES	water flow	○	○	○	○	○	○	○	●	○	○	○	○	○	○	●	○	○	○	○	○	◑	○	○	○
	low-viscosity organic liquids	○	○	○	○	◑	◑	◑	●	○	○	○	○	○	(3) ○	●	○	○	(2) ○	○	○	◑	○	○	○
	high-viscosity organic liquids	(4) ○	●	●	◑	●	●	○	●	◑	○	○	○	◑	●	●	●	○	(2) ○	◑	◑	●	◑	◑	(5) ○
	gases at pressure near atmospheric	(5) ○	○	(5) ○	◑	○	○	◑	○	○	●	○	○	○	○	○	(15) ●	●	●	◑	◑	○	●	●	●
	gases at high pressure	○	○	○	○	○	○	○	○	(13) ◑	●	○	○	○	○	○	(15) ●	●	●	◑	◑	○	●	◑	◑
	hot liquid (T > 200 °C)	○	○	○	○	○	○	○	●	◑	○	◑	◑	●	○	●	●	○	●	◑	◑	●	●	◑	(8) ○
	hot gases and steam (T > 200 °C)	○	○	○	○	○	○	○	○	◑	●	●	●	◑	○	○	(15) ●	●	●	●	●	◑	●	◑	●
	suspension of solids in liquids	(1) ◑	○	●	○	●	●	○	●	●	○	●	●	●	◑	●	●	○	○	●	○	◑	●	◑	○
	two-phases flow : liq + vap ; liq + gas	(1) ◑	◑	◑	◑	●	◑	○	●	●	●	●	●	●	●	●	●	●	(7) ○	●	●	●	●	○	○
		O	V	N	E	P	AP	TR	L	R	CP	RV	ND	TU	VXS	VXP	FO	NMR	EM	TOF	DF	TH	WB	AM	CA
FLUID-DYNAMIC FEATURES	low Reynolds numbers (Re ≤ 2000, laminar flow regime)	●	●	●	●	◑	◑	○	○	◑	◑	○	○	●	●	●	●	○	○	○	○	(14) ○	●	○	○
	2000 ≤ Reynolds ≤ 6000 (transition flow regime)	(4) ○	●	●	●	●	◑	○	○	◑	◑	○	○	●	●	●	○	○	○	○	○	●	●	○	○
	very small liquid flows	(5) ◑	○	●	◑	(9) ○	(9) ○	●	◑	○	●	○	○	◑	●	●	○	○	○	●	●	○	○	◑	○
	very small gas flows	●	(5) ◑	●	●	(9) ◑	(9) ◑	●	○	○	●	○	○	●	●	●	(15) ●	●	●	●	●	○	●	◑	●
	very wide rangeability	●	●	●	●	●	●	●	–	○	○	(10) ○	(10) ○	○	○	○	○	○	○	○	○	○	○	○	○
	pulsating flow	(9) ◑	(9) ◑	(9) ◑	(9) ◑	●	●	◑	○	◑	●	●	●	○	○	●	●	◑	○	○	●	○	●	●	●

		O	V	N	E	P	AP	TR	L	R	CP	RV	ND	TU	VXS	VXP	FO	NMR	EM	TOF	DF	TH	WB	AM	CA
PLANT FEATURES	very large water pipes	(5) ◑	○	●	○	○	○	(6) ●	●	●	◑	●	●	(6) ●	(6) ●	●	●	●	◑	○	○	●	●	●	●
	very large air or gas ducts	(5) ◑	○	●	◑	○	○	●	●	●	●	●	●	(6) ●	(6) ●	●	●	●	●	●	●	(6) ●	●	●	●
	low pressure losses	●	○	●	○	○	○	(6) ●	●	○	●	◑	◑	(6) ◑	(6) ●	●	●	○	○	○	○	○	●	●	●
	long life without maintenance or recalibration	○	○	○	○	○	○	◑	○	◑	●	●	●	●	○	○	◑	○	(12) ○	○	○	○	○	◑	○
		O	V	N	E	P	AP	TR	L	R	CP	RV	ND	TU	VXS	VXP	FO	NMR	EM	TOF	DF	TH	WB	AM	CA
PERFORMANCES FEATURES	high accuracy measurement of liquid flowrate	◑	◑	◑	●	●	●	●	○	◑	●	○	○	○	○	●	◑	◑	○	◑	◑	●	○	○	○
	high accuracy measurement of liquid quantity	●	●	●	●	●	●	●	●	●	●	○	○	○	○	○	◑	◑	○	○	○	●	○	○	○
	high accuracy measurement of gas flowrate and quantity	◑	◑	◑	●	●	●	◑	○	●	●	○	○	○	○	◑	●	●	●	●	●	◑	●	(11) ◑	(11) ◑

TABLE 11

NOTES in Table 11

(1) Eccentric orifice

(2) Conductive fluids

(3) The flow range decreases as the dynamic viscosity rises

(4) Eccentric and conical entrance orifices

(5) Depending on the pressure losses

(6) For insertion type: ○

(7) For high concentration at the secondary phase: ●

(8) A calibration to the measuring fluid temperature is required

(9) Depends on the secondary elements

(10) The flow range for gas flows is wider than the flow range for liquid flows

(11) Only for high pressure (heavy gas)

(12) A periodic cleaning of the electrodes is required; there are flowmeters with self-cleaning electrodes

(13) Only for metal tube type

(14) Only for laminar element version

(15) The existing fluidic oscillator flowmeters are limited to liquid service, even if theoretically the Coanda effect is suitable also to measure gas flows

KEY

○ generally suitable, or very useful in certain circumstances

◑ worth considering, or sometimes suitable

● unsuitable, or not normally applicable

For flowmeter types, see "ABREVIATIONS" in Table 1

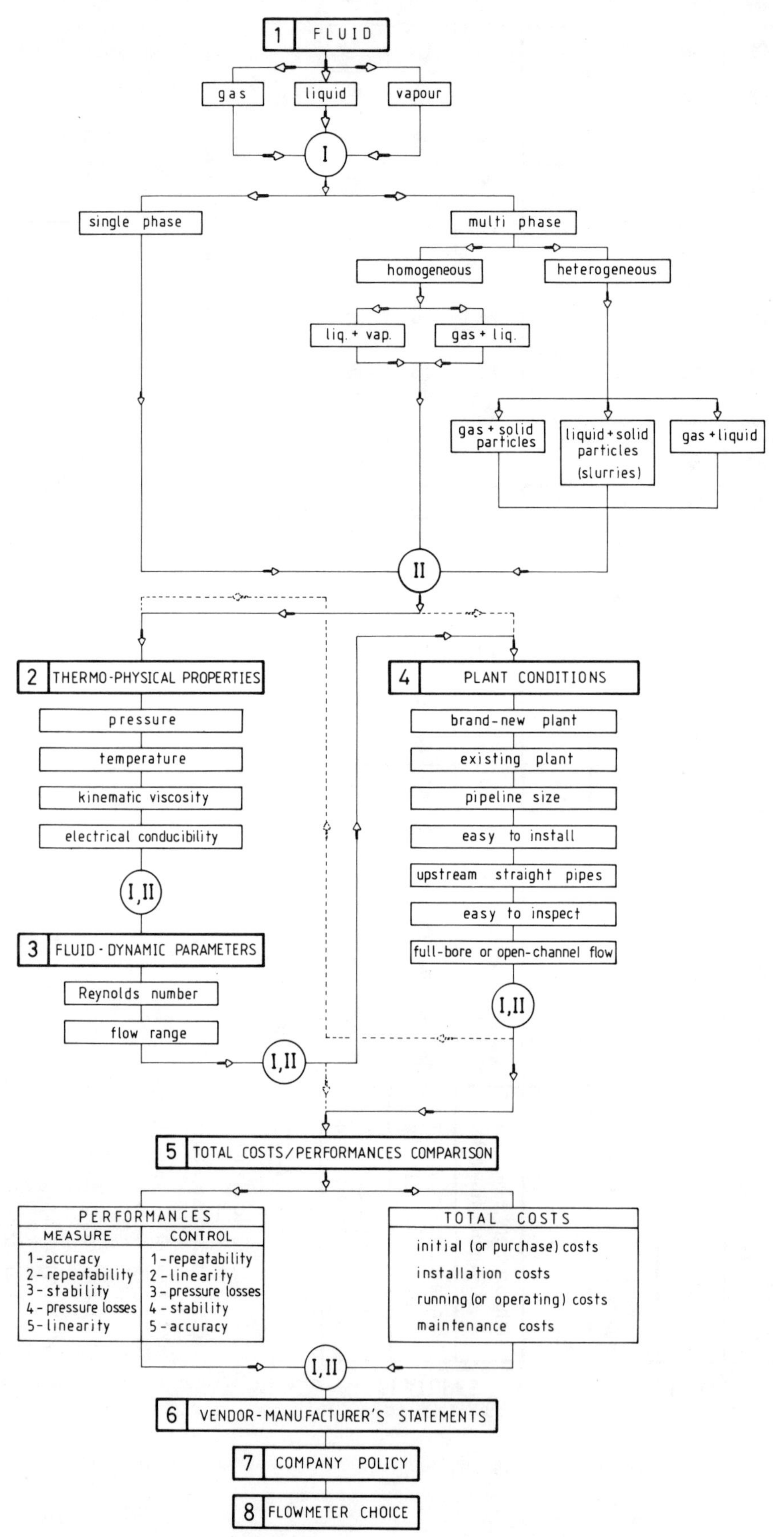

Figure 2—Flow chart showing link between Reynolds number and rangeability.

cannot be removed by the strainers of the degasifier. As with multiphase homogeneous fluids, the choice of "probable" flowmeters for heterogeneous fluids is limited by the unsuitability of most intrusive methods. This is due to the problem of erosion, to the accumulation of materials on surfaces, and to the plugging of apertures and the retarding of moving parts. Target meters (and sometimes vortex-shedding meters) can be used with heterogeneous fluids and owe a measure of their success to such adaptability.

However, independent of the fluid's phase (single, multiphase) or nature (homogeneous, heterogeneous), all the methods based on classic fluid-dynamic equations are unsuitable when the fluid itself is non-Newtonian. Such fluids can be measured by non-intrusive electromagnetic and ultrasonic flowmeters, which nevertheless require great care because of unidentified velocity profiles. Therefore, the only recommended meters are positive displacement devices (which "pack" the fluid quite apart from its motion) and, even better, Coriolis acceleration types (which can be defined "non-intrusive true mass flowmeters" because of their mode of operation).

The second step in the selection, as shown in Figure 2, is the analysis of the thermophysical properties of the fluid to be measured. Thus, the flowmeters that can be damaged or made ineffective by some thermophysical properties are eliminated from the list of "probable" flowmeters so far drawn up. For instance, high temperature or pressure values damage some meters such as variable area, turbine, and positive displacement, while high kinematic viscosity makes some other meters ineffective, such as turbine and vortex. Finally, low electrical conductivity of the fluid makes electromagnetic meters completely useless.

Next, after consideration of the interaction between the meter and its plant, two further selections can be made, leading to steps 3 (fluid-dynamic parameters) and 4 (plant conditions) and resulting in a new list of "feasible" flowmeters.

As regards step 3, it is crucial to determine the Reynolds number, which is a dimensionless parameter characterizing the type of flow regime (laminar, turbulent, or transition) and the consequent velocity profiles (respectively, parabolic, quasi one-dimensional, and power law). Only the electromagnetic and Coriolis acceleration meters always work correctly in any of the above flow conditions. Most meters are designed for use at high Reynolds numbers (> 6000) since they are based on the one-dimensional hypothesis (fully turbulent velocity profile). The use of each flowmeter is limited by a stated minimum Reynolds number, below which correction factors may be used or special calibrations or compensations effected at the low end of the scale to avoid a drop of the metrological performances, which are strongly influenced by the transition flow. Nevertheless, ultrasonic meters can work accurately, when calibrated, both with low and high Reynolds numbers, i.e., in turbulent and laminar flow zones, while they are error-prone in the transition zone. Positive displacement meters for liquids are usually designed for good accuracy at low Reynolds numbers; units for gas service are more accurate at high Reynolds numbers. Vortex-shedding meters are strongly influenced by a minimum Reynolds number because they work only when the relationship between the volumetric flow rate and the vortex-shedding frequency is linear. Therefore, they are not recommended for Reynolds numbers below 10,000. In fact, such a relationship is linear for Reynolds numbers above 10,000, while it is nonlinear from 2,000 to 10,000, even though vortices are shed.

The Reynolds number, moreover, affects the accuracy of all methods that infer volumetric flow rate from the measurement of a point velocity, such as the Pitot tube and all insertion meters [34].

Each flowmeter, as previously stated, operates within a given flow range, but the wider the range the greater the need to check the minimum and maximum Reynolds numbers for best performance of the meter. The flow chart in Figure 2 shows the tight link between Reynolds number and rangeability, which follows it immediately. In the case of wide flow ranges it is inappropriate to use square root scale methods, such as the head meters (orifice, Venturi tube, nozzle, elbow, Pitot tube, averaging Pitot tube) since their typical flow range is 4:1. Laminar flowmeters, of course, are not included in the above list, as they are based on linear scale methods. Target meters, however, though based on square root scale, can widen the assessed flow range up to 10:1, thanks to the great versatility of their secondary units. A flow range ratio similar to the one above is characteristic of all other linear scale methods. In the case of ultrasonic, vortex-shedding, and positive displacement meters, the ratio may be doubled (20:1).

Moreover, the study of the interaction between the flowmeter and its plant (step 4) is crucial for the choice of the right meter, especially when the instruments were not provided for in the original design of the plant. Therefore, a dotted line in the flow chart in Figure 2 means that sometimes step 4 can immediately follow step 1 and be followed in its turn by steps 3 and 2.

Before installing a flowmeter, it is necessary to take into account the size of the pipe and the possible absence of long straight pipes. If the size of the pipe is large, some meters must be left out of the original list of "probable" flowmeters, and the more economical, though less accurate, insertion-type meters are often preferred, also owing to the cheapness of their assemblage in pre-existing plants. Moreover, in pre-existing plants, many meters may not be installed correctly because of the absence of straight pipes, which, as shown in Table 1, are required by the majority of flowmeters based on the one-dimensional hypothesis (i.e., fully turbulent flow).

Furthermore, it is also necessary to consider quick and economical maintenance and, of course, easy access to the meter itself prior to installation.

The above considerations (steps 1 to 4) lead to a list of "feasible" flowmeters, which the instrumentation specialist compares on the basis of the ratio of total costs and peformances (step 5), thus drawing up a final graded list. The order of preference in the graded list must be determined by the evaluation of the aims of the measurement and of the consequent performance expected from the meter, as well as by the evaluation of the total costs related with the chosen flowmeter. In fact, the meter can be used either for metering or for control, and, therefore, the priorities in typical performance criteria are different in the two cases, as shown in Figure 2. For example, accuracy must be put in the first rank mostly in custody transfer applications, while, in many applications, performance is not as crucial when a meter is used for control because it may be difficult to evaluate the influence of flowmeter accuracy on plant efficiency or on the quality of the final product. Finally, repeatability and reliability have priority in the choice of a flowmeter for industrial process plants.

Some considerations must also be made about the total cost of a flowmeter (cost is a critical factor in the selection of any equipment), with consideration given to initial purchase, installation, operating, and maintenance costs [70]. In Figure 3, the approximate initial costs of the most popular flowmeters are shown as a function of pipeline sizes; these curves should be used only as a general guide to illustrate trends and to obtain a price comparison. Some considerations that will help to evaluate the hardware installation costs can be inferred from Table 1. Hints on mounting time are given in the "Ease of Installation" column (Table 1), though the real cost of manpower is liable to variations due to location and the working situation. Most meters must be mounted between flanges (wafer), thus involving very high costs for large pipe sizes. Therefore, the much more economical, though less accurate, insertion-type meter may be preferred with large diameters.

Running or operating costs are strictly related to electrical energy costs, i.e., with pumping costs (Figure 4), which are in linear proportion with the pressure losses induced by the meter at a given flow. With regard to Figure 4 it should be also pointed out that the pumping costs are connected to the plant working day fraction (i.e., the working hours in a year). The mutual position of each flowmeter curve in Figure 4 should have the same trend for different pipe sizes.

Finally, maintenance costs depend on the nature of the operation and on the time lag between operations. Furthermore, the time lag is a function of the performance and the efficiency required by the plant. In the end, it is hardly worth remembering that any consideration of cost must always be kept up to date, especially in any long-term evaluation program.

The order of preference in the graded list that arises from the comparison on the basis of the ratio of total costs to performances must be compared with the vendor-manufacturers' and users' statements (step 6) about previous experiences in similar situations, provided that the source of information is reliable. Either the vendor-manufacturers' statements or considerations concerning the general policy of the company (step 7) can eventually overthrow the stated order of preference.

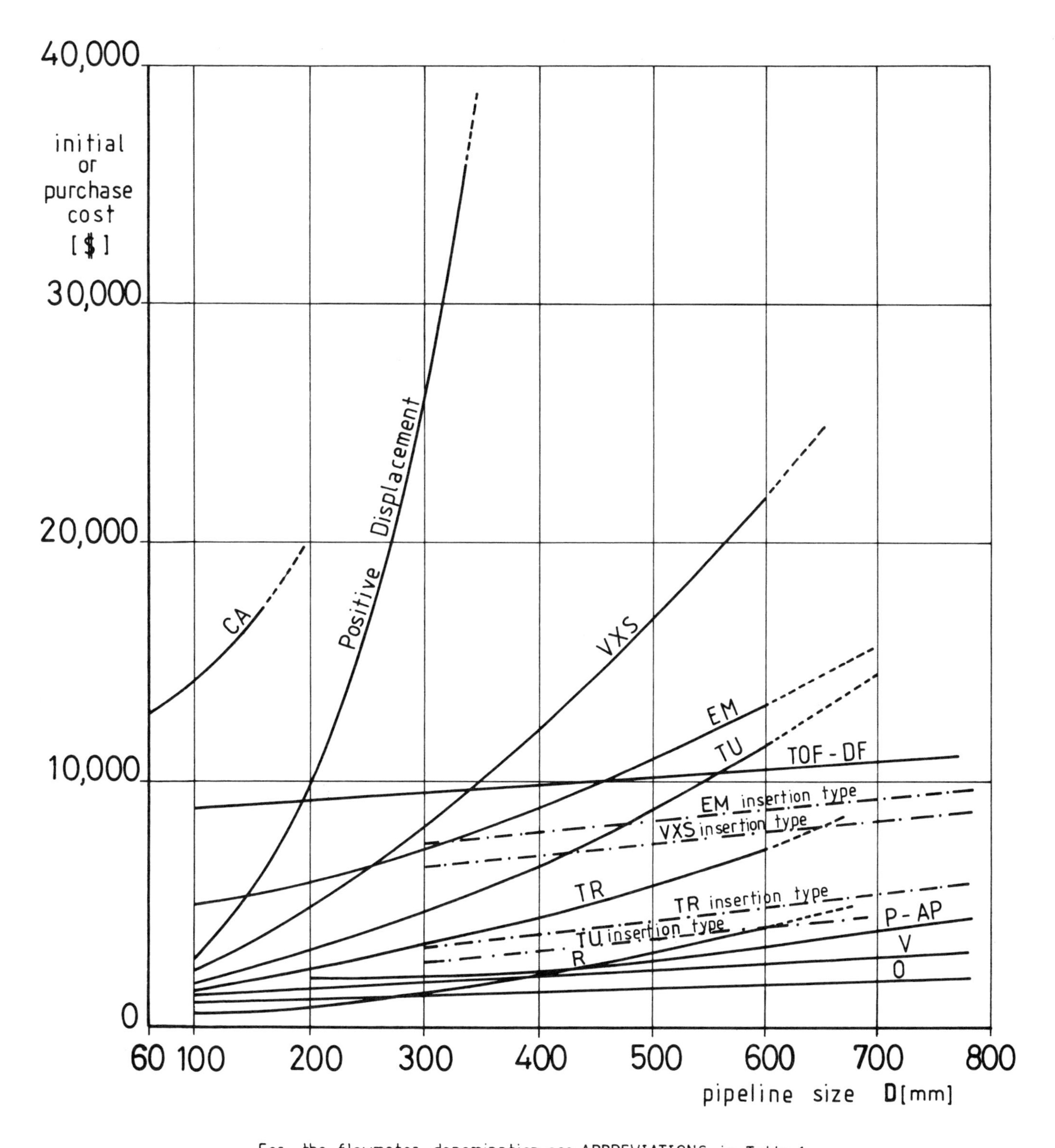

For the flowmeter denomination see ABBREVIATIONS in Table 1

Figure 3—Initial cost of the most popular flowmeters

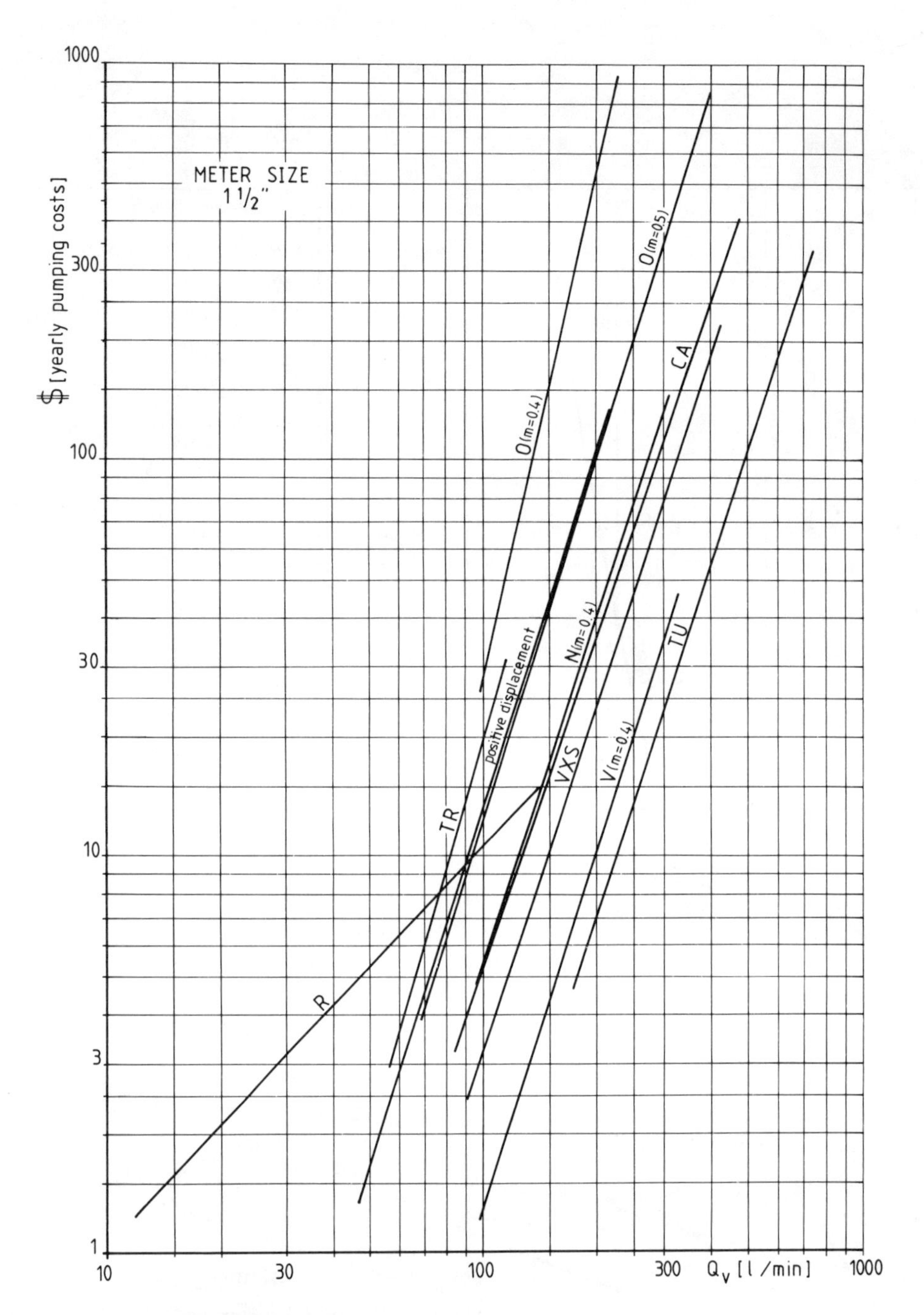

Figure 4—Yearly pumping costs of the most popular flowmeters

NOTES:

For flowmeter types, see "ABREVIATIONS" in Table 1

$$\text{Yearly pumping costs} = K \frac{Q_V \Delta p C}{\eta} \text{wdf}$$

K = units conversion factor = $1.47 \cdot 10^2$ kWh/1 mbar

Q_v =volumetric flow rate [l/min]

Δ_p = pressure losses [mbar]

C = cost of electrical energy = 0.10 \$/kWh

η = pump/compressor - efficiency factor = 0.75

wdf = working day fraction = 300/360 = 83.3% = 0.833

m = throat area ratio = d^2/D^2

REFERENCES

[1] E.A. Spencer: "The influence of the flow on flowmeter selection," *Proceedings* of the First Symposium on Flow: Its Measurement and Control in Science and Industry, Pittsburgh, 1971.

[2] J. Hall: "Flowmeters—matching applications and devices," *Instruments & Control Systems*, Feb. 1978, pp 17-22.

[3] A.T.J. Hayward: "A comprehensive survey of flowmeter types," *Processing*, May 1981, pp 23-37.

[4] A.T.J. Hayward: *Flowmeters, A Basic Guide and Source-Book for Users*, The Macmillan Press Ltd, London, 1979.

[5] F. Cascetta, P. Vigo: "Scelta opportuna del misuratore di portata: la diversificazione rende piu' complessa la decisione," *Automazione e Stumentazione*, Feb. 1986, pp 159-168.

[6] J.I. Sackett: "Measurement of thermal-hydraulic parameters in liquid-metal-cooled fast breeder reactors," *Proceedings* of The International Centre for Heat and Mass Transfer: Measurement Techniques in Heat and Mass Transfer, Hemisphere Publishing Corporation, New York, 1985.

[7] A. Aschenbrenner, N. Watanabe: "Intercomparison of gas flow test facilities in the United States of America, Europe and Japan," *Proceedings* of the International Conference of Flow Meaurement - Flowmeko '85, Melbourne, Aug. 1985, pp 114-122.

[8] R.L. Moore: *Measurement Fundamentals*, Vol. 1, Instrument Society of America, Research Triangle Park, 1982.

[9] R.W. Miller: *Flow Measurement Engineering Handbook*, McGraw-Hill Book Company, New York, 1983.

[10] D.W. Spitzer: *Industrial Flow Measurement*, Instrument Society of America, Research Triangle Park, 1984.

[11] J.P. DeCarlo: *Fundamentals of Flow Measurement*, Instrument Society of America, Research Triangle Park, 1984.

[12] V. Cavaseno: "Flowmeter choices widen," *Chemical Engineering*, Jan. 1978, pp 55-57.

[13] P. Harrison: "Flow measurement. A state of the art review," *Chemical Engineering*, Jan. 1980, pp 97-104.

[14] G.L. Grebe: "Flowmeter selection: broadening - but still a challenge," *InTech*, Oct. 1984, pp 65-69.

[15] T.J.S. Brain, R.W.W. Scott: "Survey of pipeline flowmeters," *J. Phys. E.: Sci. Instrum.*, Vol. 15, 1982, pp 967-980.

[16] A. Krigman: "Flowmeter measurement: a state of flux," *InTech*, Oct. 1984, pp 9-13.

[17] J. Hall: "Flow computers: what's happening?" *I & CS - The Industrial and Process Control Magazine*, Feb. 1985, pp 69-76.

[18] R.S. Medlock: "The techniques of flow measurement," *Measurement and Control*, part 1: Dec. 1982, pp 458-463; part 2: Jan. 1983, pp 9-13.

[19] F. Cascetta, P. Vigo: "La misura della portata di massa tramite l'effetto Coriolis," *Proceeding* of 41° National Congress A.T.I., Naples, Sep. 1986, pp VI-21, VI-32, to be published in *La Termotecnica* in 1987.

[20] D.J. Thomas: "Selecting the right flowmeter - Part I: The six favorites," *InTech*, May 1977, pp 55-62.

[21] W.S. Corcoran, J. Honeywell: "Practical methods for measuring flows," *Chemical Engineering*, Jul. 1975, pp 86-92.

[22] *The Flowmeter Industry: A Strategic Analysis*, Venture Development Corporation, Wellesley, 1981.

[23] D.J. Lomas: "Selecting the right flowmeter - Part II: Comparing candidates," *InTech*, Jun. 1977, pp 71-77.

[24] A.T.J. Hayward: "Flowmeter choice," *Engineering*, Apr. 1977, pp 307-310.

[25] A.T.J. Hayward: "Choose the flowmeter right for the job," *Processing*, Mar. 1980, pp 47-52.

[26] A.T.J. Hayward: "Which flowmeter for which job?" *Processing*, Dec. 1982, pp 35-43.

[27] S. Bailey: "Tradeoffs complicate decisions in selecting flowmeters," *Control Engineering*, Apr. 1980, pp 75-79.

[28] Yang Gen-Sheng, Zhang Bao-Xin: "The effect of the configuration and installing conditions of the pressure taps of an orifice plate on the flow measurement," *Proceedings* of The Second Symposium of Flow: Its Measurement and Control in Science and Industry, St. Louis, ISA Pub., 1981, pp 387-398.

[29] G.E. Mattingly: "Improving flow measurement performance: Research techniques and prospects," *InTech*, Jan. 1985, pp 57-64.

[30] E.A. Spencer: "Progress on international standardization of orifice plates for flow measurement," *Int. J. Heat & Fluid Flow*, Vol. 3, No. 2, June. 1982, pp 59-66.

[31] ISO 5167 - "Measurement of Fluid Flow by Means of Orifice Plates, Nozzles, and Venturi Tubes

Inserted in Circular Cross-Section Conduits Running Full," 1980.

[32] J.W. Murdock, C.J. Foltz, C. Gregory: "Performance characteristics of elbow flowmeters," *Trans. ASME, Journal of Basic Engineering*, Vol. 86-3, Sep. 1984, p. 498.

[33] ISO 3966 - "Measurement of Fluid Flow in Closed Conduits - Velocity Area Method Using Pitot Static Tubes," 1977.

[34] ISO 7145 - "Determination of Flowrate of Fluids Inclosed Conduits of Circular Cross-Section - Method of Velocity Measurement at One Point of the Cross-Section," 1982.

[35] W.H. Hickman: "Annubar properties investigation," *Proceedings* of ISA Industry-Oriented Conference and Exhibit, Milwaukee, WI, 1975.

[36] N.Q. Thoi, W. K. Soh: "On the averaging technique and the discharge coefficient of the Annubar averaging flow sensor," *ISA Transactions*, Vol. 18, No. 1, 1979, p. 41.

[37] M. Stepler: "Drag-body flowmeter," *Instruments and Control Systems*, Nov. 1962, p. 97.

[38] D.E. Curran: "Laboratory determination of flow coefficient values for the target flowmeter at low Reynolds number flow," *Proceedings* of The Second Symposium of Flow: Its Measurement and Control in Science and Industry, St. Louis, ISA Pub., 1981, pp 263-276.

[39] C. Janniello, P. Vigo: "Alcune note sulla misura della resistivita' acustica dei materiali porosi," *Rivista Italiana di Acustica*, Vol. VII, 4, 1983, pp 255-258.

[40] J. Levis: "Rotameters," *Measurement and Control*, Sep. 1980, p. 170.

[41] *Fluid Meters - Their Theory and Application*, American Society of Mechanical Engineers, New York, Sixth Edition, 1971.

[42] G.H. Stine: "The development of the turbine flowmeter," *ISA Transactions*, Vol. 16, No. 3, p. 17.

[43] H. J. Evans; "Turbine flowmeter for gases," *Instruments and Control Systems*, Mar. 1964, p. 103.

[44] RP31.1, "Specification, Installation and Calibration of Turbine Flowmeters," Instrument Society of America, 1977.

[45] E.D. Woodring: "Magnetic turbine flowmeters," *Instruments and Control Systems*, June. 1969, p. 133.

[46] S.J. Bailey: "Two new flowmeters have no moving parts," *Control Engineering*, Dec. 1969, p. 73.

[47] H. Yamasaki, S. Honda: "A unified approach to hydrodynamic oscillator type flowmeters," *Proceedings* of The Second Symposium of Flow: Its Measurement and Control in Science and Industry, St. Louis, ISA Pub., 1981, pp 191-198.

[48] D.F. White, A.E. Rodeley, C.L. McMurtie: "The vortex shedding flowmeter," *Proceedings* of The First Symposium of Flow: Its Measurement and Control in Science and Industry, Pittsburgh, ISA Pub., 1971, pp 967-974.

[49] R.B. Beale, M.T. Lawler: "Development of a wall-attachment fluidic oscillator applied to volume flow metering," *Proceedings* of The First Symposium of Flow: Its Measurement and Control in Science and Industry, Pittsburgh, ISA Pub., 1971, pp 989-996.

[50] H.H. Dijstelbergen: "The performance of a swirl flowmeter," *Journal of Physics E: Sci. Instrum.*, Vol. 3, 1970, pp 886-888.

[51] P.H. Herzl: "The system approach to high performance gas flow measurement with the swirlmeter," *Proceedings* of The First Symposium of Flow: Its Measurement and Control in Science and Industry, Pittsburgh, ISA Pub., 1971, pp 963-966.

[52] L.E. Drain: *The Laser Doppler Technique*, J. Wiley & Sons Pub., London, 1980.

[53] R.P. Keech, J. Coulthard: "Advances in cross-correlation flow measurement and its application," *Proceedings* of Int. Conf. on Flow Measurement - Flomeko '85, Melbourne, Aug. 1985, pp 195-202.

[54] W.K. Genthe, W.R. Vander Heyden, J.H. Battocletti, W.S. McCormick, H.M. Snowball: "NMR applied to flow measurement, *InTech*, Nov. 1986, pp 53-58.

[55] V. Cushing: "Electromagnetic flowmeter," *The Review of Scientific Instruments*, Aug. 1965, p. 1142.

[56] J. Hemp, M.L. Sanderson: "Electromagnetic flowmeters - a state of the art review," *Proceedings* of the Int. Conf. on Advances in Flow Measurement Techniques, Coventry, England, Sep. 1981, pp 319-340.

[57] L.C. Lynnworth: "Ultrasonic flowmeter - eight types," *Measurement and Control*, Sep. 1980, p. 126.

[58] M.L. Sanderson, J. Hemp: "Ultrasonic flowmeters - a review of the state of the art," *Proceedings* of the

Int. Conf. on Advances in Flow Measurement Techniques, Coventry, England, Sep. 1981, pp 157-178.

[59] T.R. Schmidt: "What you should know about clamp-on ultrasonic flowmeters," *InTech*, May 1981, pp 59-62.

[60] H.M. Morris: "Ultrasonic flowmeter uses wide beam technique to measure flow," *Control Engineering*, Jul. 1980, pp 99-101.

[61] H.W. Stoll: "The measurement of flow with density compensation as applied to liquids and gases," *ISA Journal*, May 1955, p. 159.

[62] J.M. Benson: "Survey of thermal devices for measuring flow," *Proceedings* of The First Symposium of Flow: Its Measurement and Control in Science and Industry, Pittsburgh, ISA Pub., 1971, pp 549-554.

[63] L.T. Akeley, L.A. Batchelden, D.S. Cleveland: "Gyro-integrating mass flowmeter," ASME Paper no. 58-IRD3, 1958.

[64] W. Masnik: "Flo-Tron: misuratore di piccole portate," *ATA*, Lug./Ago. 1978, pp 3-6.

[65] K.O. Plache: "Coriolis/gyroscopic flowmeter," *Mechanical Engineering*, Mar. 1979, pp 36-41.

[66] J.P. Tullis, J. Smith: "Coriolis mass flowmeter," Paper No. 6.3 presented to the Fluid Mechanics Silver Jubilee Conference, Glasgow, 1979.

[67] E. Dahlin, A. Young, R. Blake, C. Guggenheim, S. Kaiser, A. Levien: "Digital precision mass flowmeter," Paper presented at ISA '84.

[68] F. Cascetta, S. della Valle, A.R. Guido, P. Vigo: "A new type of coriolis acceleration mass flowmeter," Paper accepted for IMEKO XI, Houston, Oct, 1988.

[69] F. Cascetta, S. della Valle, A.R. Guido, P. Vigo: "A Coriolis mass flowmeter based on a new type of elastic suspension," 21° International Conference on Control of Industrial Processes, Milan, Italy, October 1987.

[70] W.M. Reese, Jr.: "Factor the energy costs of flow metering into your decisions," *InTech*, Jul. 1980, pp 36-38.